Modelling and Machine Learning Methods for Bioinformatics and Data Science Applications

Modelling and Machine Learning Methods for Bioinformatics and Data Science Applications

Editors

Monica Bianchini
Maria Lucia Sampoli

MDPI • Basel • Beijing • Wuhan • Barcelona • Belgrade • Manchester • Tokyo • Cluj • Tianjin

Editors
Monica Bianchini
University of Siena
Italy

Maria Lucia Sampoli
University of Siena
Italy

Editorial Office
MDPI
St. Alban-Anlage 66
4052 Basel, Switzerland

This is a reprint of articles from the Special Issue published online in the open access journal *Mathematics* (ISSN 2227-7390) (available at: https://www.mdpi.com/journal/mathematics/special_issues/Math_Model_Machine_Learning_Bioinformatics_Data_Science).

For citation purposes, cite each article independently as indicated on the article page online and as indicated below:

LastName, A.A.; LastName, B.B.; LastName, C.C. Article Title. *Journal Name* **Year**, *Volume Number*, Page Range.

ISBN 978-3-0365-2840-3 (Hbk)
ISBN 978-3-0365-2841-0 (PDF)

© 2021 by the authors. Articles in this book are Open Access and distributed under the Creative Commons Attribution (CC BY) license, which allows users to download, copy and build upon published articles, as long as the author and publisher are properly credited, which ensures maximum dissemination and a wider impact of our publications.

The book as a whole is distributed by MDPI under the terms and conditions of the Creative Commons license CC BY-NC-ND.

Contents

About the Editors . . . . **vii**

Preface to "Modelling and Machine Learning Methods for Bioinformatics and Data Science Applications" . . . . **ix**

Md Al Masum Bhuiyan, Ramanjit K. Sahi, Md Romyull Islam, Suhail Mahmud
Machine Learning Techniques Applied to Predict Tropospheric Ozone in a Semi-Arid Climate Region
Reprinted from: *Mathematics* **2021**, *9*, 2901, doi:10.3390/math9222901 . . . . **1**

Cecilia Berardo, Iulia Martina Bulai and Ezio Venturino
Interactions Obtained from Basic Mechanistic Principles: Prey Herds and Predators
Reprinted from: *Mathematics* **2021**, *9*, 2555, doi:10.3390/math9202555 . . . . **15**

Gerardo Alfonso Perez and Javier Caballero Villarraso
Alzheimer Identification through DNA Methylation and Artificial Intelligence Techniques
Reprinted from: *Mathematics* **2021**, *9*, 2482, doi:10.3390/math9192482 . . . . **33**

Giuseppe Alessio D'Inverno, Sara Brunetti, Maria Lucia Sampoli, Dafin Fior Muresanu, Alessandra Rufa and Monica Bianchini
Visual Sequential Search Test Analysis: An Algorithmic Approach
Reprinted from: *Mathematics* **2021**, *9*, 2952, doi:10.3390/math9222952 . . . . **47**

Niccolò Pancino, Caterina Graziani, Veronica Lachi, Maria Lucia Sampoli, Emanuel Ștefănescu, Monica Bianchini, Giovanna Maria Dimitri
A Mixed Statistical and Machine Learning Approach for the Analysis of Multimodal Trail Making Test Data
Reprinted from: *Mathematics* **2021**, *9*, 3159, doi:10.3390/math9243159 . . . . **61**

Giorgio Ciano, Paolo Andreini, Tommaso Mazzierli, Monica Bianchini and Franco Scarselli
A Multi-Stage GAN for Multi-Organ Chest X-ray Image Generation and Segmentation
Reprinted from: *Mathematics* **2021**, *9*, 2896, doi:10.3390/math9222896 . . . . **75**

About the Editors

Monica Bianchini, Ph.D., is currently an Associate Professor at the Department of Information Engineering and Mathematics of the University of Siena (Full Professor Abilitation). She received the Laurea cum laude in Mathematics and a Ph.D. degree in Computer Science from the University of Florence, Italy, in 1989 and 1995, respectively. After receiving the Laurea, for two years, she was involved in a joint project of Bull HN Italia and the Department of Mathematics (University of Florence), aimed at designing parallel software for solving differential equations. From 1992 to 1998, she was a Ph.D. student and a Postdoc Fellow with the Computer Science Department of the University of Florence. Since 1999, she has been with the University of Siena. Her main research interests are in the field of machine learning, with emphasis on neural networks for structured data and deep learning, approximation theory, information retrieval, bioinformatics, and image processing. Monica Bianchini has authored more than one hundred papers and has been the editor of books and Special Issues in international journals in her research field. She has been a participant in many research projects focused on machine learning and pattern recognition, founded by both Italian Ministry of Education (MIUR) and University of Siena (PAR scheme), and she has been involved in the organization of several scientific events, including the NATO Advanced Workshop on Limitations and Future Trends in Neural Computation (2001), the 8th AI*IA Conference (2002), GIRPR 2012, the 25th International Symposium on Logic-Based Program Synthesis and Transformation, and the ACM International Conference on Computing Frontiers 2017. Prof. Bianchini served as an Associate Editor for *IEEE Transactions on Neural Networks* (2003–09), *Neurocomputing* (from 2002), *Int. J. of Computers in Healthcare* (from 2010), and *Frontiers in Genomics* (section Computational Genomics). She is a permanent member of the Editorial Board of *IJCNN, ICANN, ICPRAM, ESANN, ANNPR*, and *KES*.

Maria Lucia Sampoli, Ph.D., is currently Associate Professor in Numerical Analysis at the Department of Information Engineering and Mathematics of the University of Siena. She graduated in Mathematics with Honors from the University of Florence in 1994, and in 1998 received a PhD in Computational Mathematics and Operative Research from the University of Milan. In the same year she received a postdoc fellowship at the Institute of Applied Geometry of the Technical University of Darmstadt (Germany). In 1999–2000 she was a CNR Senior Fellow at Department of Mathematics of the University of Siena, where she initially was a Research Assistant and later in 2002 became a tenured Assistant Professor in Numerical Analysis. In 2010 she joined the Department of Information Engineering and Mathematics. Her research interests are mainly focused on the use of splines functions in various applications, from shape preserving interpolation and approximation to numerical approximation of elliptic problems (Isogeometric Analysis), quasi-interpolations techniques, Numerical quadrature, and Phythagorean curves and surfaces. Maria Lucia Sampoli has been invited speaker (as a plenary or as a speaker at mini-symposia) at several international conferences as well as involved in the organization of various conferences and workshops. She has been involved in many research projects including as a coordinator for some of them.

Preface to "Modelling and Machine Learning Methods for Bioinformatics and Data Science Applications"

With the enormous amount of data flowing from a variety of real-world problems, Artificial Intelligence (AI), and in particular Machine Learning (ML) and Deep Learning (DL) techniques, have powered new achievements in many complex applications, that were prohibitive with deterministic approaches. These advances, which are based on a multidisciplinary research framework involving computer science, numerical analysis and statistics, come from research efforts in both industry and academia and are particularly well suited to address complex problems in data science and, more specifically, in biotechnological and medical applications. While these methods have proven to be astounding in performance, they still suffer from a sort of opacity, meaning that their produced results, though correct and quickly obtained, are difficult to be interpreted or explained, a fundamental drawback especially for those problems where people's lives are at stake. Therefore, a deeper understanding of the fundamental principles of machine and deep learning methods is mandatory in order to evidence both their advantages and limitations. Though a variety of different approaches exists to face the problem of explainability—ranging from methods that try to understand on which features an AI model bases its prediction, to the construction of ad hoc architectures which some logical knowledge can be extracted from—we think that a viable alternative is that of a *contamination* between mathematical modeling and machine learning, in the belief that the insertion of the equations deriving from the physical world in the data-driven models can greatly enrich the information content of the sampled data, allowing us to simulate very complex phenomena, with drastically reduced calculation times and interpretable solutions. The application of such hybrid techniques to structured data, such as time series or graphs, however, opens to other interesting challenges, aimed at determining whether these techniques are able to generalize to different problems, how much the data structure (for instance, sampling from a subspace or manifold) affects the method, and how to choose appropriate (hyper-)parameters to ensure a good fit, while still avoiding overfitting.

This Special Issue brings together researchers from different disciplinary fields, who focus on building theoretical foundations and presenting cutting-edge applications for deep learning applications. We hope this issue will serve as a hint for researchers to reflect on fundamental issues in a wide variety of fields, including pure and applied mathematics, statistics, computer science and engineering, to join forces to integrate different approaches suitable for solving complex problems, quickly, reliably and understandably for human experts.

The contributions collected can be divided into two main categories, relating, respectively, to the application of mathematical models/DL techniques to the study of biological macrosystems and to the automatic analysis/prediction of medical data for the prognosis of human diseases.

For the first category, the paper titled "Machine Learning Techniques Applied to Predict Tropospheric Ozone in a Semi-Arid Climate Region" describes a comparative evaluation of a large class of statistical modeling methods for classifying high or low ozone concentration levels. Indeed, ground-level ozone exposure has led to a significant increase in environmental risks, since it adversely affects not only human health but also some delicate plants and vegetation.

Additionally, in the paper "Interactions Obtained from Basic Mechanistic Principles: Prey Herds and Predators", four different predator–prey–herd models are presented, that are derived assuming that the prey gathers in herds, that the predator can be specialist—i.e., it feeds on only one species—or generalist—i.e., it feeds on multiple resources—and considering two functional responses, the herd-linear and herd-Holling type II functional responses. The paper aims at deriving their mathematical formulation from the individual-level state transitions, and compare the models' dynamics in terms of equilibria, stability and bifurcation diagrams. The predator–prey–herd antagonistic behavior has been widely observed in population ecology, especially in aquatic species and insects, and has been proven to deeply affect niche expansion and speciation.

To the second category belongs the manuscript "Alzheimer Identification through DNA Methylation and Artificial Intelligence Techniques", which presents a nonlinear approach for identifying combinations of CpG DNA methylation data as biomarkers for Alzheimer disease (AD). Indeed, the possibility of having techniques that can determine earlier if an individual has AD is becoming increasingly important, especially after the FDA approval of the first drug for AD treatment (there were drugs before it targeting some of the effects of the illness, but not the actual illness itself). Such an early diagnosis will be possible soon, thanks to non-invasive medical tests to capture methylation data, simply based on blood.

Two contributions, namely "Visual Sequential Search Test Analysis: An Algorithmic Approach" and "A Mixed Statistical and Machine Learning Approach for the Analysis of Multimodal Trail Making Test Data", are devoted to the automatic analysis of Trail Making Test (TMT) data. TMT is a popular neuropsychological test, commonly used in clinical settings as a diagnostic tool for the evaluation of some frontal functions, that provides qualitative information on high order mental activities, including speed of processing, mental flexibility, visual spatial orientation, working memory and executive functions. Such data are preprocessed in the form of sequences and treated with an algorithmic approach based on the episode matching method, or in the form of scan-path images, that can be processed via DL and clustering methods, for distinguishing patients affected by the extrapyramidal syndrome and by chronic pain from healthy subjects. A statistical analysis, based on the blinking rate and on the pupil size, is also carried out, to help classifying different pathologies.

Finally, the paper "A Multi-Stage GAN for Multi-Organ Chest X-ray Image Generation and Segmentation" proposes a deep learning approach to the generation of realistic synthetic images—particularly useful in medical applications where the scarcity of data often prevents the use of DL architectures—that can be employed to train a segmentation network. Segmentation is, in fact, the preventive step for automatic image analysis and classification, and has proven fundamental, for instance, in order to diagnose COVID-19 based on lung damage.

Monica Bianchini, Maria Lucia Sampoli

Editors

Article

Machine Learning Techniques Applied to Predict Tropospheric Ozone in a Semi-Arid Climate Region

Md Al Masum Bhuiyan [1,*], Ramanjit K. Sahi [1], Md Romyull Islam [1] and Suhail Mahmud [2]

1. Department of Mathematics & Statistics, Austin Peay State University, Clarksville, TN 37044, USA; sahir@apsu.edu (R.K.S.); mislam@my.apsu.edu (M.R.I.)
2. Earth & Environmental Systems Institute (EESI), The Pennsylvania State University, State College, PA 16802, USA; sfm6095@psu.edu
* Correspondence: bhuiyanm@apsu.edu; Tel.: +1-931-221-7964

Abstract: In the last decade, ground-level ozone exposure has led to a significant increase in environmental and health risks. Thus, it is essential to measure and monitor atmospheric ozone concentration levels. Specifically, recent improvements in machine learning (ML) processes, based on statistical modeling, have provided a better approach to solving these risks. In this study, we compare Naive Bayes, K-Nearest Neighbors, Decision Tree, Stochastic Gradient Descent, and Extreme Gradient Boosting (XGBoost) algorithms and their ensemble technique to classify ground-level ozone concentration in the El Paso-Juarez area. As El Paso-Juarez is a non-attainment city, the concentrations of several air pollutants and meteorological parameters were analyzed. We found that the ensemble (soft voting classifier) of algorithms used in this paper provide high classification accuracy (94.55%) for the ozone dataset. Furthermore, variables that are highly responsible for the high ozone concentration such as Nitrogen Oxide (NOx), Wind Speed and Gust, and Solar radiation have been discovered.

Keywords: tropospheric ozone; machine learning; El Paso-Juarez; semi-arid climate

1. Introduction

Environmental problems, especially air pollution, are gaining attention as it is one of the most crucial health hazards to humans. It is an invisible killer that takes numerous human lives every year. Thus, it is essential to predict whether a day will be polluted or not. Presently, there are various pollutants in the atmosphere. Ground-level ozone especially affects human health and some delicate plants and vegetation adversely. It has been noted that high concentrations of ground-level ozone are of significant concern for many metropolitan cities in US and Mexico. In our paper, we are focusing on the border cities of El Paso in Texas and Juarez in Mexico. The climate of this region is arid and has characteristics of the urban southwestern US climate [1]. The region's air quality problem is partially the result of industrial activities and high automobile emissions in the region. Moreover, the geopolitical region of El Paso-Juarez is characterized by exceptional meteorological conditions, such as higher planetary boundary layer heights (PBLHs) than any other surrounding city, due to its complex topography.

El Paso, being a semi-arid climate region, experiences high ozone episodes in the summer season. Days with an 8-h ozone concentration of more than 70 parts per billion volume (ppbv) are defined as the High Ozone episodes [2,3]. The following Figure 1 is a representation of the annual high ozone events recorded by the Texas Commission on Environmental Quality (TCEQ) ground stations known as Continuous Ambient Monitoring Stations (CAMS) from 2000 to 2019. In this region, the highest ozone levels are commonly recorded during the summer months of June to August (Figure 2). High Ozone caused by several reasons such as high degree of temperature (June and July are the peak summer months with an average temperature of 40 degree Celsius) with calms winds (mean value of 4–5 m/s), low relative humidity and high solar radiation (roughly 1.5 Langley/min on average) [4–6].

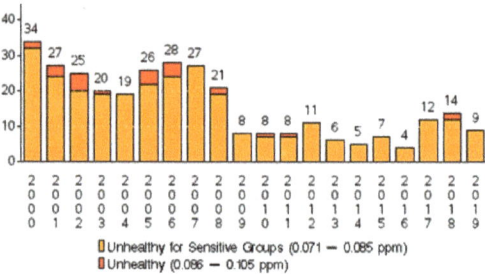

Figure 1. Total number of days with 8-hr daily ozone exceedance (above 70 ppbv) during the years of 2000–2019 in El Paso, Texas.

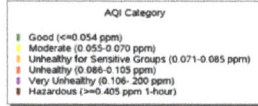

Figure 2. Ozone daily Air Quality Index (AQI) values from 2013 to 2019 for all the CAMS stations in El Paso, Texas.

A number of studies have been carried out in order to better understand the chemical and physical processes that cause high ozone concentrations in this study area [7–10]. Those studies were mainly focused on chemical composition analysis [11,12], the sources, and the physical characteristics [13] of the ozone episodes. Moreover, the topography of the study area is complex resulting in a forecast and prediction accuracy of air quality models that are less adequate to predict pollutant accurately [14]. In this paper, we propose a study of the use of machine learning techniques in classifying high or low ozone concentration levels, where highly concentrated ozone(>70 ppb) days are class 1 and low concentration ozone(<70 ppb) days are class 0. The dataset used in this paper was retrieved from TCEQ ground stations CAMS. It contains various meteorological and air pollutants that affect ground-level ozone concentration in the area of our study.

First, this dataset is filled up for further processing and then split into the training set and the testing set. In this paper, machine learning algorithms, such as Naive Bayes, K-Nearest Neighbors, Decision Tree, Stochastic Gradient Descent, and XGBoost, are utilized to predict ground ozone level concentration, and their respective accuracy scores are examined. Based on the results, the XGBoost algorithm scored the highest accuracy with 94.09 percent. The present study evaluates the first predictions of ground ozone using relatively simple predicting techniques. It can be a useful reference for scientists and meteorologists for better forecasting. In addition, the direct comparison of six different algorithms gives machine learning researchers insight into which algorithm produces the most accurate results.

This paper is constructed as follows: Section 2 describes the machine learning approaches of the six supervised algorithms. Section 3 presents the data description. The exploratory data analysis is presented in Section 4 with a focus on combining datasets. Section 5 describes the numerical experiments carried out on the real-world environment dataset. Conclusions and potential future directions will be discussed in Section 6.

2. Machine Learning Approach

This section discusses the machine learning techniques such as: Naive Bayes, K-Nearest Neighbors, Decision Tree, Stochastic Gradient Descent, and XGBoost, which were used to classify the ozone data. For this purpose, 70% of the data were applied for training a predictive model, and 30% of the data were used for testing the model based on their tuning parameters.

2.1. Naive Bayes Classifier

Naive Bayes (NB) is a supervised machine learning method, which takes the concept of conditional probability, to classify the target variable. The assumption of Bayes' theorem is to compute the probability of any class of target variable occurring based on the probability of other features that have already happened. The logic behind the model is using Bayes' theorem with the assumption of Naive, which has independent features. The model finds the probability of each class of target variables depending on some independent features. The highest probability determines the final class of target variable. The algorithm is pretty fast and is mostly used for discrete data. However, for the continuous data, the NB needs additional assumptions regarding the distribution of features. In this study, we used Gaussian NB for our continuous data (see [15] for more details).

2.2. k-Nearest Neighbors Algorithm

k-Nearest Neighbor or k-NN is a simple, non-parametric, easy to implement machine learning algorithm based on the supervised learning method. It assumes that each data point is a part of its nearest group or class based on the Euclidean distance metric. So an unclassified data point can be assigned to a class by finding what class is nearest to it. In this case, k is the hyper-parameter (typically, an odd number), that is, the number of nearest data points from unclassified data point.

2.3. Decision Tree Classifier

Decision Tree (DT) Classifiers are supervised machine learning methods that are among the easiest and inexpensive methods to grasp intuitively, due to the incorporation of decision point logic, similar to a flowchart. The inputs can be numeric or categorical. It contains nodes and edges, starting with the root node at a decision point. Edges are one of the possible answers to the question asked by the 'node'. The records are split until they cannot be split any further. So the classifier process ends when the leaf node is reached, which represents the output value for the prediction. If a separation cannot create a perfect split between the categorical response variable then it is called "impure". One metric that evaluates the impurity of a split is called "gini".

Gini impurities are calculated by subtracting from one the probability of one category squared minus the probability of another category squared and so on, as well as for any combination of the categories. These values are calculated for each node and are then aggregated as a weighted average. For numeric data, the values are sorted, and then averages are created using adjacent values and then gini impurities are derived using these. The least impure average is used as a greater than or less than split. Ranked data can be used to create splits as well. A split is chosen if its impurity is the lowest. This process is repeated to make additional splits based on the previous split [16].

2.4. XGBoost Algorithm

Gradient Boosting (GB) is a supervised algorithm that is used to predict a target variable via an ensemble technique. Chen and Guestrin proposed the Extreme GB (XGboost) algorithm on the basis of the Gradient Boosting Decision Tree (GBDT) structure [17]. In regular GB, the loss function is used only for the base model (e.g., decision tree) rather than the overall model. As compared to other algorithms, XGboost provides more information about the gradient direction and achieves a minimum of loss function quickly. The reason is that XGboost uses the second order derivative for gradient approximation [18]. This is a very fast algorithm and has several tuning parameters to improve the model, such as drop rate, sub-sample, ntrees, skip drop, and so forth.

2.5. Stochastic Gradient Descent

Gradient descent is an optimization algorithm used in finding the parameters by minimizing a cost function. The algorithm is usually slow for large datasets. The stochastic gradient descent (SGD) updates the parameters for each training data point. So the frequent updates make the algorithm computationally more expensive compared to the batch gradient descent. However, the SGD algorithm provides with the detailed information for model improvement. A brief overview of the gradient descent algorithm is as follows:

Assume that a hypothesis is $h_w(x) = w_0 + w_1 x + \cdots + \cdots$, where the parameters are $w_0, w_1, \cdots$. At this point, the cost function is:

$$J(w_0, w_1, \cdots) = \frac{1}{2m} \sum_{i=1}^{m} (h_w(x^{(i)}) - y^{(i)})^2. \quad (1)$$

The goal is to minimize the cost function $J(w_0, w_1, \cdots)$ and to find the optimum parameters $w_0, w_1, \cdots$. To do so, we use the gradient descent algorithm:

$$w_j = w_j - \alpha \frac{\partial}{\partial w_j} J(w_0, w_1, \cdots), \quad (2)$$

where $j = 0, 1, \cdots$, and α is the learning rate, or how rapidly do we want to move towards the minimum. At this point, a learning rate is specified that controls the amount of change of coefficients on each update. Overshoot can also be done for the case of large α [19].

2.6. Ensemble Methods

The Ensemble method aggregates the predictions of several similar or conceptually different models built with a given learning algorithm, improving robustness compared to a single model. Voting Classifier is one of the basic ensemble techniques. This approach works like an electoral system, in which a prediction of a new data point is made based on the machine learning models studied. In this study, we used a soft voting system where each classifier's specific weight is set with the "weight" parameter. As a probability estimate for the loss function of the SGD classifier ('hinge') is not available, we used the weights of the other classifiers. The weights were chosen based on the accuracy and other evaluation metrics. We then multiplied the probabilities and weights of each classifier and took the average of all samples. The maximum average probability is used for the final classification.

3. Data Background

Meteorological and Air pollutant parameters from six different locations of the El Paso area are used to create the database (Figure 3). Those locations represent all four different parts of the area: Urban, Rural, Industrial and Semi-industrial. El Paso possesses the typical southwestern US urban climate, which is warm and arid with poor air quality, partly due to industrial activity along the US/Mexico border, as well as unique meteorological conditions caused by local geography [20–22]. There are approximately 0.7 million people in El Paso, with 1.3 million in Cuidad Juarez, Mexico, immediately adjacent to El Paso (Padgett and Thomas, 2001). El Paso is often found to be in non-attainment of the US National Ambient Air Quality Standards (NAAQS) for O_3, CO, PM_{10}, and $PM_{2.5}$. This non-attainment is due to the combination of a high population concentration, industrial influences, and weather conditions [23].

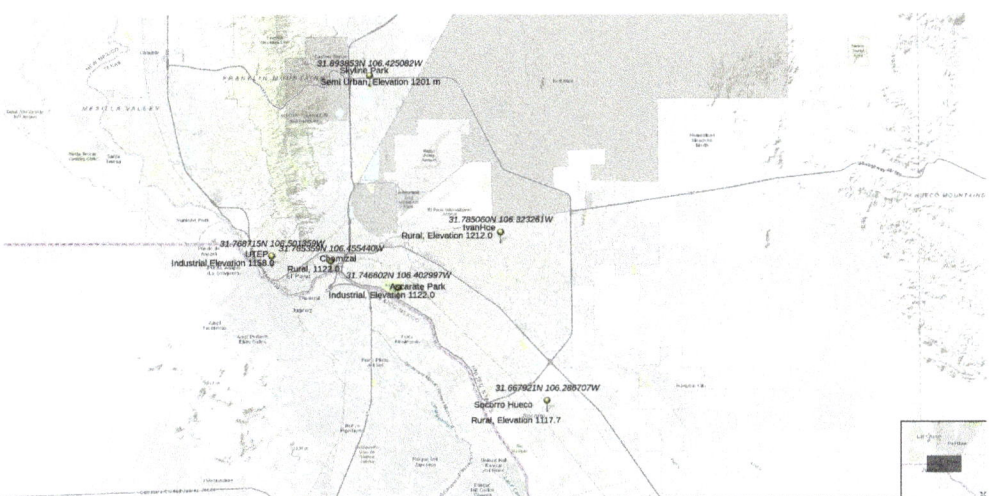

Figure 3. Six ozone monitoring stations in El Paso and their geographic centroid.

To capture the full topographic and meteorological characteristics of this study area, data from six different stations were used. Detailed information about these monitoring stations is given below in the following Table 1.

Table 1. Ozone monitoring stations.

Station	Lat/Lon	Elevation	Type
C12 UTEP	31.7682/−106.5012	1158.0	Industrial
C37 Ascarate Park	31.7467/−106.4028	1122.0	Industrial
C41 Chamizal	31.7656/−106.45522	1122.0	Rural
C49 Socorro Hueco	31.6675/−106.2880	1117.7	Rural
C72 Skyline Park	31.8939/−106.4258	1201.0	Semi Urban
C414 Ivanhoe	31.7857/−106.3235	1212.0	Rural

4. Exploratory Data Analysis

In this study, data are collected from CAMS administered by TCEQ. An average of ground-level Ozone concentrations from the six stations of monitoring stations in El Paso is provided.

As can be interpreted from Figure 4, most of the high ozone days occurred in the summer season. Minimum values of ozone days took place in January as the temperature has a positive correlation with ozone concentration. From the month of May–August, high ozone days are ubiquitous in our area of study with exceeding values from the NAAQS standard.

In Figure 5, a correlation matrix for all the different variables with ozone is presented. The relationship between Ozone and Temperature is positively correlated and ranging the value around 0.60. Meanwhile, the relation between Relative Humidity, Solar Radiation with Ozone is negatively correlated as the value ranges around (−0.5) to (−0.6).

Table 2 represents the descriptive statistics of independent variables, which includes mean, median, standard deviation, minimum values, first quartile, second quartile, third quartile and maximum values. All the fractional values are rounded to two digits after the decimal point.

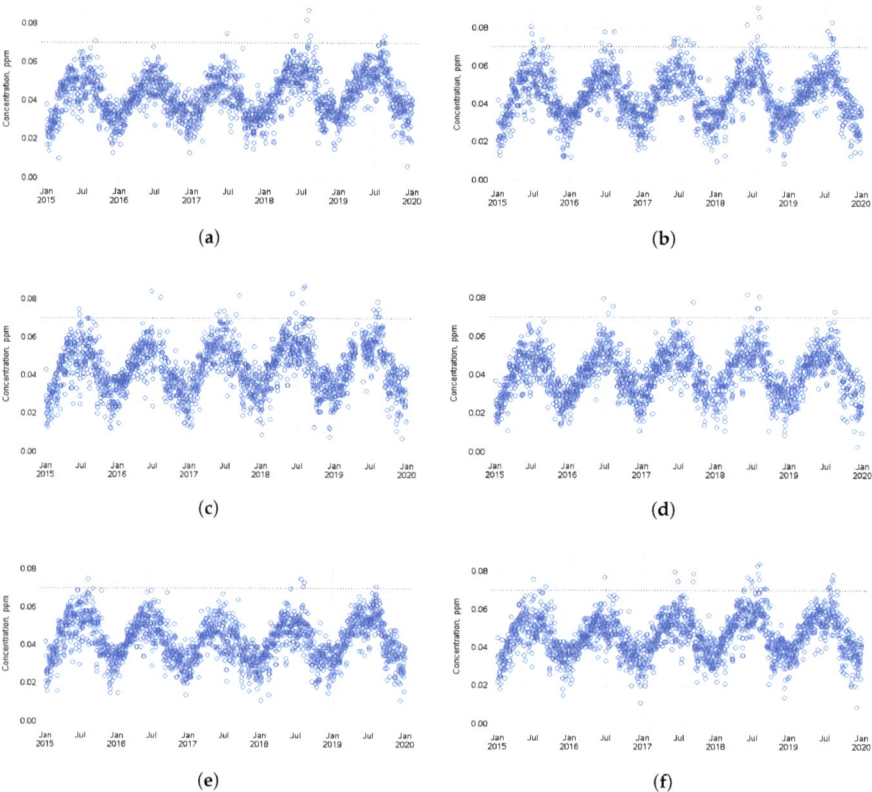

Figure 4. Daily (maximum 8 h) ozone concentration from all the stations of the study area, (**a**) Ivanhoe, (**b**) El Paso (UTEP), (**c**) Chamizal, (**d**) Ascarate Park, (**e**) Socorro Hueco, and (**f**) Skyline park.

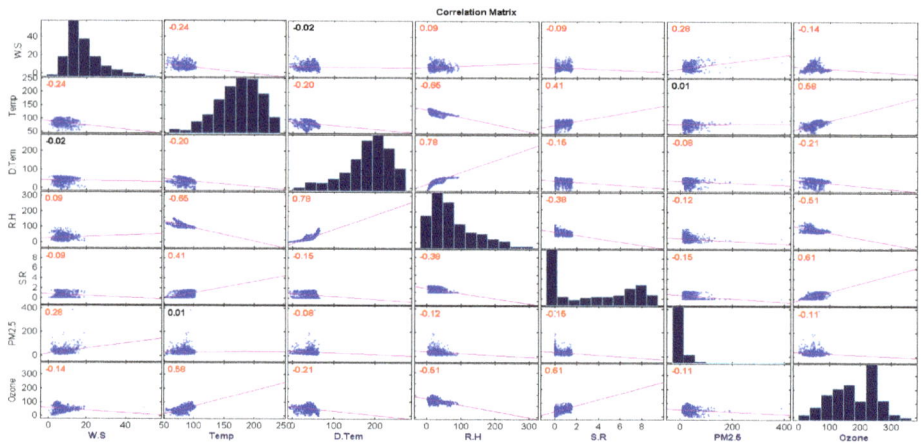

Figure 5. Correlation matrix for the dataset.

Table 2. Descriptive statistics of data.

Features	Mean	Median	STD	Min	1st Q	2nd Q	3rd Q	Max
Nitric Oxide	0.93	0.40	2.68	−1.1	0	0.4	0.8	31.8
Nitrogen Dioxide	8.66	6.70	7.07	−1.1	4.45	6.7	10.5	54.8
Oxides of Nitrogen	9.43	7.00	8.85	−1.6	4.6	7	11.1	75.1
Wind Speed	5.76	5.10	3.88	1.6	4.2	5.1	6.4	68
Resultant Wind Speed	4.41	3.80	3.7	0.2	2.5	3.8	5.6	79
Resultant Wind Direction	151.07	126.00	84.79	0	96	126	196.5	360
Maximum Wind Gust	13.48	12.60	5.78	3.2	10.5	12.6	14.9	76
Std. Dev. Wind Direction	37.97	37.00	17.3	10	21	37	51	87.4
Outdoor Temperature	89.73	91.10	8.04	32	85.25	91.1	95.2	107.1
Dew Point Temperature	42.92	42.80	9.26	16.8	36.25	42.8	48.95	89.4
Relative Humidity	22.16	18.90	13.38	0	13.2	18.9	24.8	86.6
Solar Radiation	0.64	0.51	1.12	0	0	0.51	1.17	20.4
PM_{10}	31.37	25.30	43.22	−5.9	19.4	25.3	34.4	906
$PM_{2.5}$	10.86	8.00	12.26	−3	5.3	8	11.2	146

5. Results & Discussion

This section will illustrate some experimental results of six machine learning techniques applied to our dataset.

5.1. Analysis of Fitted Models

We first analyzed the NB Classifier to create a predictive model using training data points. The NB Classifier can be highly scaled with the predictors and it requires a low number of data points. It is observed that NB is robust to the outlier and irrelevant features of data. We then used the k-NN algorithm that worked based on the parameter k. The k-NN classifier is very sensitive to the low values of k, leading to high variance and low bias of the model. On the other hand, if we take the large values of k, the model leads to low variance and high bias. Using cross-validation, we found that $k = 5$ fits well with our trained data. The decision tree algorithm does not require any scaling or normalization of our trained data. We also used the gini index to identify the important variables. Using the XGBoost algorithm, we obtained 94.09% accuracy on the test data, which is the highest accuracy among the algorithms. In this case, the in-built Lasso and Ridge Regression regularization were used to prevent the model from overfitting. The algorithm also comes with a built-in capability to handle missing values. In the SGD algorithm, we updated the weights based on each training example, not the batch as a whole. Hence, it updates the parameters for each training example one by one. The SGD approach reaches convergence much faster than batch gradient descent since it updates weights using a single data point in each iteration. So the curve of cost versus epoch for SGD algorithm is not smooth. The predictive results are shown in Tables 7 and 8 and Figures 7 and 8.

5.2. Feature Selection

The feature importance enables the machine learning algorithm to train faster, reduces the computational cost (and time) of training, and makes it simpler to interpret. It also reduces the variance of the model and improves the accuracy if the right subset is chosen. In this paper, we used different measures to select the right features (see Tables 3 and 4 and Figure 6). At this point, the ensemble model does not provide the coefficients of features due to the NB classifier. Thus, we present the coefficients of the nearest accurate model, that is, XGBoost.

Table 3. Coefficients for important feature using XGboost.

Features	Coefficients
Nitric Oxide	0.077
Nitrogen Dioxide	0.037
Oxides of Nitrogen	0.043
Wind Speed	0.037
Maximum Wind Gust	0.117
Std. Dev. Wind Direction	0.096
Outdoor Temperature	0.110
Dew Point Temperature	0.036
Relative Humidity	0.041
Solar Radiation	0.307

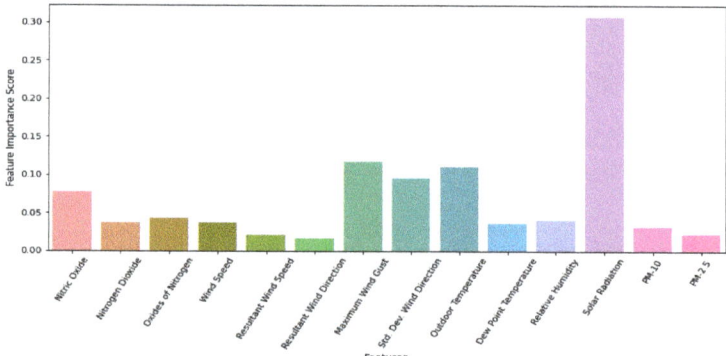

Figure 6. Visual representation of feature importance.

Table 4. Analysis of feature importance using p-values.

Features	p-Values
Nitric Oxide	6.01×10^{-1}
Nitrogen Dioxide	1.59×10^{-2}
Oxides of Nitrogen	2.85×10^{-1}
Wind Speed	5.22×10^{-12}
Resultant Wind Speed	2.50×10^{-2}
Maximum Wind Gust	1.32×10^{-4}
Std. Dev. Wind Direction	1.86×10^{-7}
Outdoor Temperature	1.12×10^{-11}
Dew Point Temperature	2.12×10^{-6}
Relative Humidity	1.07×10^{-2}
Solar Radiation	1.30×10^{-30}
PM_{10}	9.85×10^{-1}
$PM_{2.5}$	3.58×10^{-1}

The significance level and the p-values of predictors obtained from the Ordinary Least Squares [24] method are also employed in this study. The significance level (0.05) is the amount of change a feature will affect towards the final output and the p-value is the hypothesis of the significance level. At this point, the null hypothesis refers to the predictors of the model that are not significant. The higher the p-value is, the less important the feature is. From Table 4, it is clear that Nitric Oxide, Oxides of Nitrogen, Resultant Wind Direction, PM_{10}, and $PM_{2.5}$ are not important predictors, as the p-values are higher than the significance level. Thus, these features do not alter the classification and can be easily removed without causing any problems. On the other hand, Solar Radiation is most significant on days that are high in ozone. However, the Ordinary Least Square usually does not provide a good estimation for high dimensional datasets, thus we performed other machine learning processes to select important features (see Table 3 and Figure 6).

5.3. Model Evaluation

Tables 5 and 6 show that the predicted Mean Squared Error and Misclassification Rate of the models are very low for the dataset.

Table 5. Model Evaluation for Ozone dataset.

Models	Mean Squared Error	Misclassification Rate
NB	0.186	0.186
KNN	0.1	0.1
DT	0.118	0.118
XG	0.059	0.059
SGD	0.141	0.141
EN	0.055	0.055

Table 6. Model Evaluation for selected features of Ozone dataset.

Models	Mean Squared Error	Misclassification Rate
NB	0.173	0.173
KNN	0.091	0.091
DT	0.132	0.132
XG	0.068	0.068
SGD	0.127	0.127
EN	0.064	0.064

In this section, we ensemble all the methods used in this paper and present their accuracy. Tables 7 and 8 and Figures 7 and 8 show the specificity, accuracy, precision, recall, and F1 score of the models.

Specificity is used to measure the predictive performance of the classification models. Specificity informs us about the proportion of actual negative cases that our model has predicted as negative. Here, the highest value of specificity is 94.26%, which means that the Ensemble model we used is good at predicting the true negatives. To fully evaluate the effectiveness of a model, we must examine both precision and recall. Precision measures the percentage of ozone labeled as high that were correctly classified. On the other hand, recall measures the percentage of actual ozone labels that were correctly classified. F1 score is the harmonic mean of precision and recall. In this study, we also used the F1 score as an evaluation metric. The reason is that the percentage of accuracy can sometimes be disproportionately skewed by a large number of actual negatives (low ozone level) results.

We plotted the ROC curve between True Positive Rate (X-axis) and False Positive Rate (Y-axis). In Figures 9 and 10, the diagonal line represents the threshold of the ROC curve. We see that the Ensemble model has good accuracy, precision, recall, F1 score, and area under the ROC curve.

Table 7. Model Evaluation metrics analysis for Ozone dataset.

Models	Specificity (%)	Accuracy (%)	Precision (%)	Recall (%)	F1 Score [0, 1]
NB	69.67	81.36	71.75	95.91	0.82
KNN	88.52	90	86.54	91.84	0.89
DT	88.52	88.18	86	87.76	0.87
XG	92.62	94.09	91.26	95.92	0.94
SGD	88.52	85.91	85.26	82.65	0.84
EN	94.26	94.55	93.0	94.90	0.94

Table 8. Evaluation metrics analysis for selected feature of Ozone dataset.

Models	Specificity (%)	Accuracy (%)	Precision (%)	Recall (%)	F1 Score [0, 1]
NB	72.13	82.72	73.44	95.92	0.83
KNN	88.52	90.1	86.79	93.87	0.90
DT	88.52	86.81	85.57	84.69	0.85
XG	93.44	93.18	91.92	92.86	0.92
SGD	88.52	87.27	85.71	85.71	0.86
EN	93.44	93.64	92	93.87	0.93

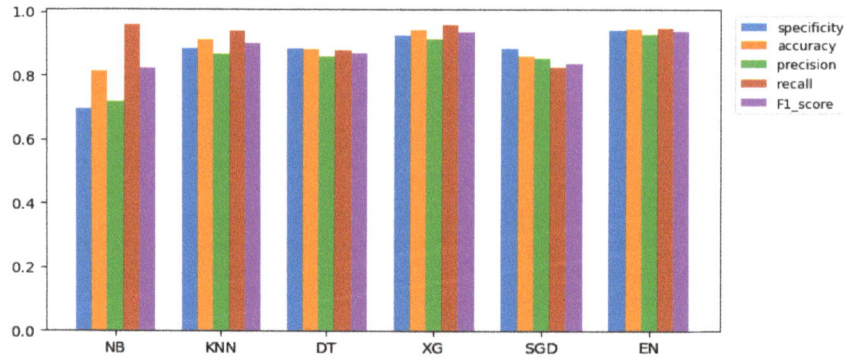

Figure 7. Model performance for Ozone dataset.

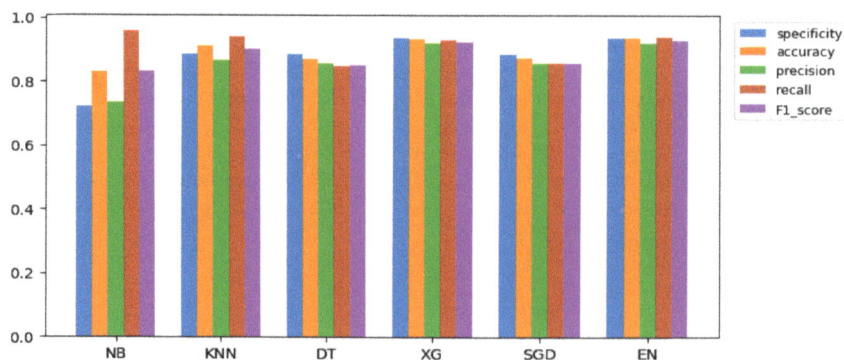

Figure 8. Model performance with selected feature for Ozone dataset.

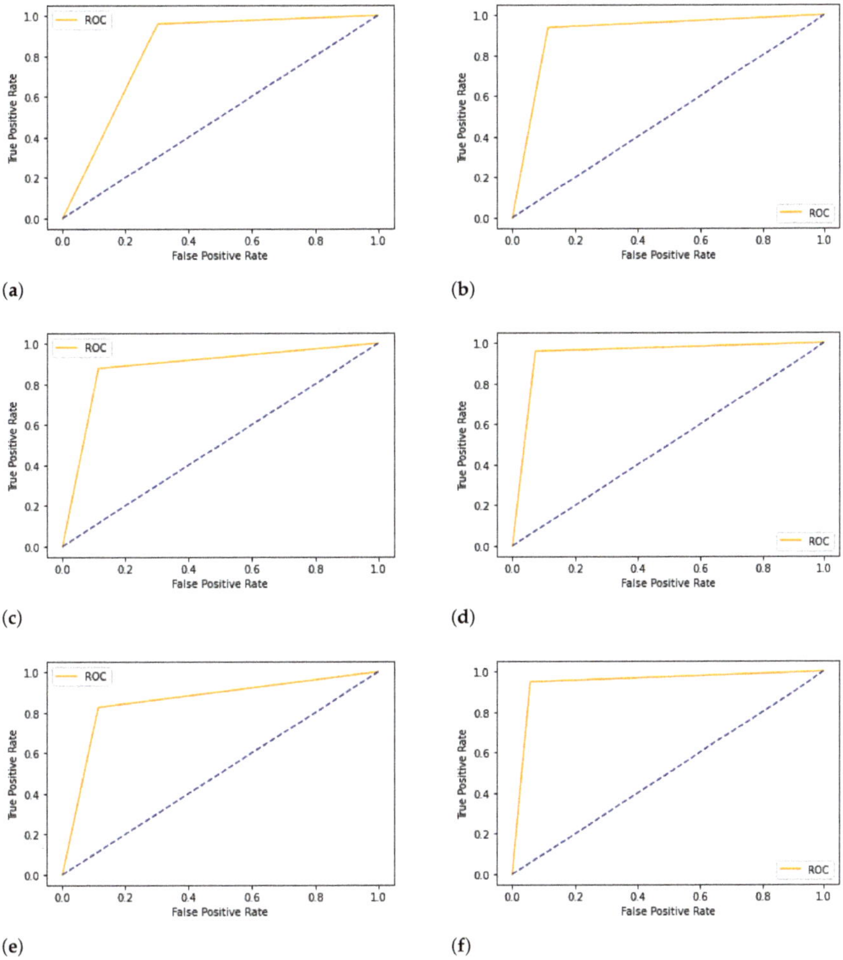

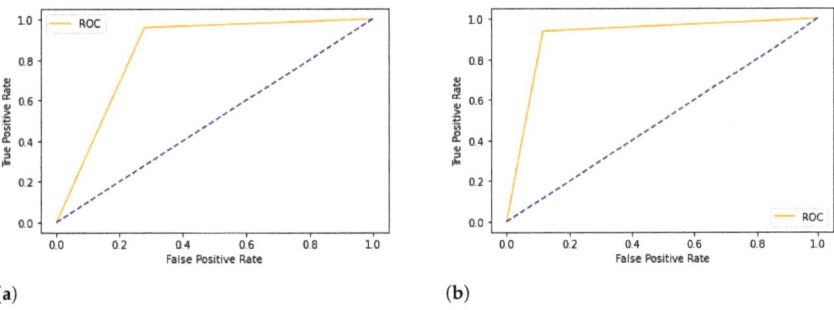

Figure 9. ROC Curve for Ozone dataset. (**a**) Fitted Naive Bayes; (**b**) Fitted K-Nearest Neighbors; (**c**) Fitted Decision Tree; (**d**) Fitted XGBoost; (**e**) Fitted Stochastic Gradient Descent; (**f**) Fitted Ensemble.

Figure 10. *Cont.*

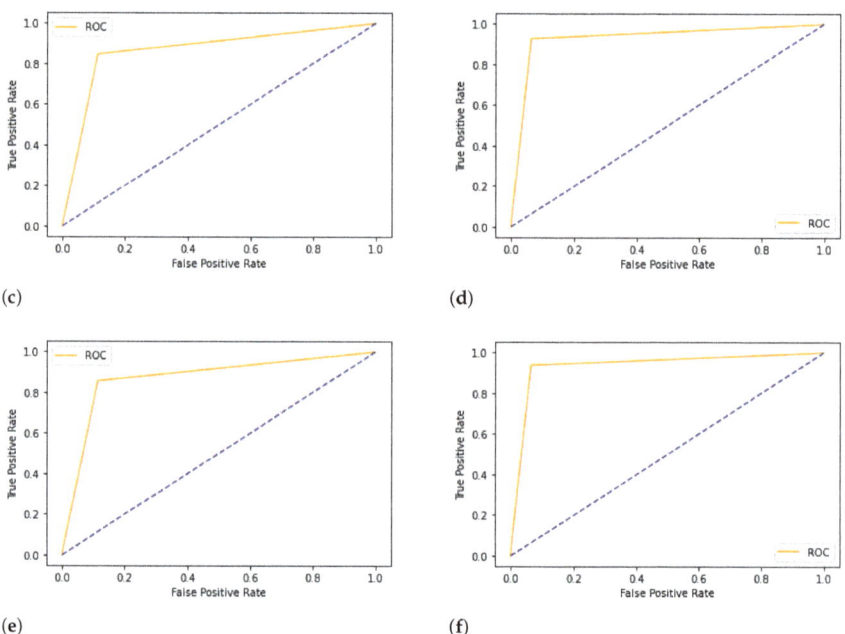

Figure 10. ROC Curve for selected featured Ozone dataset. (**a**) Fitted Naive Bayes; (**b**) Fitted K-Nearest Neighbors; (**c**) Fitted Decision Tree; (**d**) Fitted XGBoost; (**e**) Fitted Stochastic Gradient Descent; (**f**) Fitted Ensemble.

6. Conclusions

The study presented here is an extended version of analyzing machine learning for the classification of ozone in the Paso del Norte region [25]. In this case, we explore six different machine learning models in the El Paso area, in order to classify the ozone concentration in a bi-national area. In comparison to previous works in this region, this work mainly focuses on the ensemble of machine learning algorithms and study area. The algorithms used in the study are Naive Bayes, K-Nearest Neighbors, Decision Tree, Stochastic Gradient Descent, and XGBoost. The Ensemble methods have also been used to classify the ozone level. The dataset contains a mix of several air pollutants and meteorological parameters.

Initially, we did the exploratory data analysis to pre-process the data and analyze the correlation of predictors. The machine learning models have been trained with training data to build a predictive model. The performance of the predictive models has been measured by the evaluation metrics such as Mean Squared Error, Misclassification Rate, Accuracy, Specificity, Precision, Recall, and F1 Score. The XGBoost model and p-values of predictors from the OLS technique are used to determine the important variables that are useful in detecting high or low ozone days.

The Ensemble Method was able to predict 94.55% of ozone observations with high or low levels. The results indicate that the high concentration of ozone is highly influenced by Nitric Oxide, Nitrogen Dioxide, Oxides of Nitrogen, Wind Speed, Maximum Wind Gust, Std. Dev. Wind Direction, Outdoor Temperature, Relative Humidity, and Solar Radiation (see Tables 3 and 4 and Figure 6).

Author Contributions: M.A.M.B. and S.M. contributed to the supervision and project administration, M.A.M.B., R.K.S., M.R.I., and S.M., contributed to the conceptualization, methodology, and result analysis. All authors have read and agreed to the published version of the manuscript.

Funding: This research was funded by Austin Peay State University.

Acknowledgments: The authors like to thank the Texas Commission of Environmental Quality and Atmospheric Physics Lab (UTEP) for all their support.

Conflicts of Interest: The authors declare no conflict of interest.

References

1. Lee, S.H.; Kim, S.W.; Angevine, W.M.; Bianco, L.; McKeen, S.A.; Senff, C.J.; Zamora, R.J. Evaluation of urban surface parameterizations in the WRF model using measurements during the Texas Air Quality Study 2006 field campaign. *Atmos. Chem. Phys.* **2011**, *11*, 2127. [CrossRef]
2. Pernak, R.; Alvarado, M.; Lonsdale, C.; Mountain, M.; Hegarty, J.; Nehrkorn, T. Forecasting Surface O3 in Texas Urban Areas Using Random Forest and Generalized Additive Models. *Aerosol Air Qual. Res.* **2019**, *19*, 2815–2826. [CrossRef]
3. Part, V. Federal Register Document. *Fed. Regist.* **1997**, *62*, 38855–38896.
4. Seila, R.L.; Main, H.H.; Arriaga, J.L.; Martnez, G. Atmospheric volatile organic compound measurements during the 1996 Paso del Norte Ozone Study. *Sci. Total. Environ.* **2001**, *276*, 153–169. [CrossRef]
5. Mahmud, S.; Karle, N.N.; Fitzgerald, R.M.; Lu, D.; Nalli, N.R.; Stockwell, W.R. Intercomparison of Sonde, WRF/CAMx and Satellite Sounder Profile Data for the Paso Del Norte Region. *Aerosol Sci. Eng.* **2020**, *4*, 277–292. [CrossRef]
6. Karle, N.N.; Mahmud, S.; Sakai, R.K.; Fitzgerald, R.M.; Morris, V.R.; Stockwell, W.R. Investigation of the Successive Ozone Episodes in the El Paso–Juarez Region in the Summer of 2017. *Atmosphere* **2020**, *11*, 532. [CrossRef]
7. Brown, M.J.; Muller, C.; Wang, G.; Costigan, K. Meteorological simulations of boundary-layer structure during the 1996 Paso del Norte Ozone Study. *Sci. Total. Environ.* **2001**, *276*, 111–133. [CrossRef]
8. Einfeld, W.; Church, H.W.; Yarbrough, J.W *Winter Season Air Pollution in El Paso-Ciudad Juarez*; United States Environmental Protection Agency, Region VI, Air, Pesticides Toxics Division: Washington, DC, USA, 1995.
9. Funk, T.H.; Chinkin, L.R.; Roberts, P.T.; Saeger, M.; Mulligan, S.; Figueroa, V.P.; Yarbrough, J. Compilation and evaluation of a Paso del Norte emission inventory. *Sci. Total. Environ.* **2001**, *276*, 135–151. [CrossRef]
10. Ordieres, J.B.; Vergara, E.P.; Capuz, R.S.; Salazar, R.E. Neural network prediction model for fine particulate matter ($PM_{2.5}$) on the US–Mexico border in El Paso (Texas) and Ciudad Juárez (Chihuahua). *Environ. Model. Softw.* **2005**, *20*, 547–559. [CrossRef]
11. Shi, C.; Fernando, H.J.S.; Yang, J. Contributors to ozone episodes in three US/Mexico border twin-cities. *Sci. Total. Environ.* **2009**, *407*, 5128–5138. [CrossRef] [PubMed]
12. Zora, J.E.; Sarnat, S.E.; Raysoni, A.U.; Johnson, B.A.; Li, W.W.; Greenwald, R.; Sarnat, J.A. Associations between urban air pollution and pediatric asthma control in El Paso, Texas. *Sci. Total. Environ.* **2013**, *448*, 56–65. [CrossRef] [PubMed]
13. Mahmud, S. Optimization of Regional Scale Numerical Weather Prediction & Air Quality Model for the Paso Del Norte Region. Ph.D. Thesis, The University of Texas at El Paso, El Paso, TX, USA, 2020.
14. Mahmud, S. *The Use of Remote Sensing Technologies and Models to Study Pollutants in the Paso del Norte Region*; The University of Texas at El Paso: El Paso, TX, USA, 2016.
15. Moraes, R.M.; Machado, L.S. Gaussian naive bayes for online training assessment in virtual reality-based simulators. *Mathw. Soft Comput.* **2009**, *16*, 123–132.
16. Pandya, R.; Pandya, J. C5. 0 algorithm to improved decision tree with feature selection and reduced error pruning. *Int. J. Comput. Appl.* **2015**, *117*, 18–21. [CrossRef]
17. Chen, T.; Guestrin, C. Xgboost: A scalable tree boosting system. In Proceedings of the 22nd ACM Sigkdd International Conference on Knowledge Discovery and Data Mining, San Francisco, CA, USA, 13–17 August 2016; pp. 785–794.
18. Quinto, B. *Next-Generation Machine Learning with Spark*; Springer: Berlin, Germany, 2020.
19. Stochastic Gradient Descent. Available online: https://en.wikipedia.org/wiki/Stochastic-gradient-descent (accessed on 23 September 2021).
20. MacDonald, C.P.; Roberts, P.T.; Main, H.H.; Dye, T.S.; Coe, D.L.; Yarbrough, J. The 1996 Paso del Norte Ozone Study: Analysis of meteorological and air quality data that influence local ozone concentrations. *Sci. Total Environ.* **2001**, *276*, 93–109. [CrossRef]
21. Mahmud, S.; Wangchuk, P.; Fitzgerald, R.; Stockwell, W. Study of Photolysis rate coefficients to improve air quality models. *Bull. Am. Phys. Soc.* **2016**, *61*, 15.
22. Fujita, E.M. Hydrocarbon source apportionment for the 1996 Paso del Norte Ozone Study. *Sci. Total. Environ.* **2001**, *276*, 171–184. [CrossRef]
23. Mahmud, S.; Bhuiyan, M.A.M.; Sarmin, N.; Elahee, S. Study of wind speed and relative humidity using stochastic technique in a semi-arid climate region. *AIMS Environ. Sci.* **2020**, *7*, 156–173. [CrossRef]
24. Pavelescu, F.M. Features Of The Ordinary Least Square (Ols) Method. Implications for the Estimation Methodology. *J. Econ. Forecast.* **2004**, *1*, 85–101.
25. Bhuiyan, M.A.M.; Mahmud, S.; Sarmin, N.; Elahee, S. A Study on Statistical Data Mining Algorithms for the Prediction of Ground-Level Ozone Concentration in the El Paso–Juarez Area. *Aerosol Sci. Eng.* **2020**, *4*, 293–305. [CrossRef]

 *mathematics*

Article

Interactions Obtained from Basic Mechanistic Principles: Prey Herds and Predators

Cecilia Berardo [1,*], Iulia Martina Bulai [2,3] and Ezio Venturino [3,4]

[1] Department of Mathematics and Statistics, University of Helsinki, FI-00014 Helsinki, Finland
[2] Department of Mathematics, Informatics and Economics, University of Basilicata, I-85100 Potenza, Italy; iulia.bulai@unibas.it
[3] GNCS Research Group, INdAM, I-00185 Rome, Italy; ezio.venturino@unito.it
[4] Department of Mathematics Giuseppe Peano, University of Torino, I-10100 Torino, Italy
* Correspondence: cecilia.berardo@helsinki.fi

Abstract: We investigate four predator–prey Rosenzweig–MacArthur models in which the prey exhibit herd behaviour and only the individuals on the edge of the herd are subjected to the predators' attacks. The key concept is the herding index, i.e., the parameter defining the characteristic shape of the herd. We derive the population equations from the individual state transitions using the mechanistic approach and time scale separation method. We consider one predator and one prey species, linear and hyperbolic responses and the occurrence of predators' intraspecific competition. For all models, we study the equilibria and their stability and we give the bifurcation analysis. We use standard numerical methods and the software Xppaut to obtain the one-parameter and two-parameter bifurcation diagrams.

Keywords: predator–prey model; herd behaviour; herd shape; linear functional response; Holling type II functional response; bifurcation analysis

1. Introduction

Modelling herd behaviour of a population with ordinary differential equations, via a spatial factor with herding index (see Laurie et al. [1]), avoids an explicit spatial representation.

Already in 1987, Liu and colleagues [2] used the herding index, represented by an exponent, to give a non-linear incidence rate in epidemiological models, and later this concept found several applications in the same context [3,4].

The herding index, α in this article, also appeared in predator–prey dynamics with prey grouped in herd. In particular, the first studies comprised only the case $\alpha = 1/2$ by assuming a two-dimensional herd representation [5,6], while later works allowed for a fixed herding index in the range $[1/2, 1)$ [6,7]. The concept of herding index was further extended to models other than ordinary differential equations, such as time delay equations, [8], and fractional differential equations [9,10].

In predator–prey models, the assumption that the prey gather in a herd has a direct effect on the shape of the functional response, as the encounter rate of the predator and prey individuals increases with the prey density according to the herding index. Therefore, in the simple case of no prey handling, we obtain the herd-linear functional response. More complex is the herd-Holling type II functional response, already used by Djilali in [8] in a time delay differential model, which we derive in this paper from a system of fast time-scale state transitions.

The mechanistic derivation of the functional response follows from the time-scale separation method that has been recently formalised by Berardo et al. [11] and previously used in the works by Geritz and Gyllenberg [12–14]. In particular, analogously to Metz and Diekmann [15], we assume the predators in two states, searching and handling, and model the state transitions on a fast time-scale compared to other processes. The equilibrium distribution of the predators between the two states depends on the density of the prey available for capture on the edge of the herd and, as a consequence, the functional response varies with the prey density in a similar way as the Holling type II.

Finally, in this paper we study four different predator–prey–herd models. All the possible combinations arise from the following situations: first of all, assuming that the prey gather in herds, secondly, assuming that the predator can be specialist, i.e., feeding only on one species, or generalist, i.e., feeding on multiple resources, and lastly, considering two functional responses, the herd-linear and herd-Holling type II functional responses. The aim of the paper is to derive their mathematical formulation from the individual-level state transitions, and compare the models' dynamics in terms of equilibria, stability and bifurcation diagrams.

The paper is organised as follows. In Section 2, we introduce the mechanistic derivation of the herd-Holling type II functional response for the predator–prey dynamics. In Section 3, we give partial results on the equilibrium and linear stability analysis, and the bifurcation diagrams obtained with numerical methods. In Section 4, we outline the main outcomes and draw our conclusions.

2. Materials and Methods

Mechanistic Derivation of the Functional Response for the Predator–Prey Dynamics

We model the scenario where a prey species x and a predator species y interact in the following way. The prey gather in herds and the predators are partitioned among searching individuals y^S and handling individuals y^H. We define the handling time as the time necessary for killing, eating and digesting the prey, as well as resting and giving birth.

Suppose that there is one prey herd and multiple predator individuals. Assume further that

(i) the geometric shape of the herd is fixed, e.g., a circle in two dimensions or a ball in three dimensions,

(ii) the density of individuals inside the shape stays constant independently of the size of the herd,

(iii) the predators can attack the prey on the edge of the herd or in a small boundary layer of the herd.

Therefore, the number of individuals exposed to predation is proportional to x^α, where the exponent α is defined as *herding index* [1] and depends only on the characteristic shape of the herd. Examples are $\alpha = \frac{1}{2}$ for a circle in two dimensions or $\alpha = \frac{2}{3}$ for a sphere in three dimensions. Intermediate values of α would model fractal or multi connected sets; such exponents are investigated in [7].

Let us assume that the attack rate for a searching predator is ax^α, i.e., the per capita attack rate is proportional to the number of available prey according to mass action. Thus, the time until prey capture becomes $\frac{1}{ax^\alpha}$. When we exclude handling, the predator functional response is linear and takes the form $f(x) = ax^\alpha$ (*herd-linear functional response*). Alternatively, let h denote the handling time, then the time between two successive catches by the same predator is $\frac{1}{ax^\alpha + h}$.

All in all, consider the following fast processes

$$\left(y^S\right) \xrightarrow{ax^\alpha} \left(y^H\right) \quad \text{the predator meets the prey and the prey is caught,} \tag{1}$$

$$\left(y^H\right) \xrightarrow{h^{-1}} \left(y^S\right) \quad \text{the predator stops handling.}$$

The above interactions were first used by Metz and Diekmann [15] to mechanistically derive the Holling type II functional response and what differs here is the exponent α in the encounter rate ax^α, defining predation on the edge of the herd. We apply the time-scale separation method between fast and slow processes (for details on this modelling approach we refer to [11]). In particular, birth and death happen on a slower time scale. By converting the individual-level processes in (1) into differential equations, considering the rates at which individuals leave and enter the two subsets y^S and y^H, the equations for the fast-time dynamics reduce to

$$\frac{dy^S}{dt} = -ax^\alpha y^S + h^{-1} y^H, \qquad (2)$$

$$\frac{dy^H}{dt} = +ax^\alpha y^S - h^{-1} y^H, \qquad (3)$$

where we recall that a is the predator's attack rate and h denotes the handling time as defined in the individual-level processes above. The total population density y is constant on the fast time scale (as y^S and y^H verify $\frac{dy}{dt} = \frac{dy^S}{dt} + \frac{dy^H}{dt} = 0$), therefore, we can reduce the system to one equation by using the conservation law $y = y^S + y^H$. By using this law and solving the equilibrium equation for y^S, i.e., setting the right hand side of (2) and (3) to zero, we obtain a unique quasi-steady state for the fast dynamics

$$y^S = \frac{y}{1 + ahx^\alpha}, \quad y^H = y - y^S. \qquad (4)$$

By definition, the corresponding functional response is given by the average number of prey caught by a searching predator per unit of time, i.e., $f(x) = \frac{ax^\alpha y^S}{y}$, and, plugging the expression for y^S at the quasi-equilibrium in (4), we obtain

$$f(x) = \frac{ax^\alpha}{1 + ahx^\alpha}. \qquad (5)$$

We name the functional response in (5) the *herd-Holling type II functional response*. Note that when $\alpha = 1$ we recover the Holling type II functional response and when $\alpha = 2$ we obtain the Holling type III functional response. However, in the present context with prey herd geometry, we necessarily have $\alpha \in (0,1)$. In this case, the shape of the functional response is similar to the Holling type II functional response as it is concave and saturating, but the behaviour near the origin is different as it is infinitely steep.

3. Results

In this section we present the theoretical and numerical results for the following the Rosenzweig and MacArthur [16] models with functional responses derived in Section 2:

(i) specialist predator and herd-linear functional response $f(x) = ax^\alpha$;
(ii) generalist predator and herd-linear functional response $f(x) = ax^\alpha$;
(iii) specialist predator and herd-Holling type II like functional response $f(x) = \frac{ax^\alpha}{1+ahx^\alpha}$;
(iv) generalist predator and herd-Holling type II like functional response $f(x) = \frac{ax^\alpha}{1+ahx^\alpha}$.

The analysis is organised as follows. First, we check the unboundedness of the prey and predator populations, we derive the dimensionless version of the equations, we compute the equilibrium points and we study their stability by applying linear stability analysis. Sections 3.1–3.4 cover this first part of the study and give the analytical results for each of the models above. Finally, in Section 3.5, we use standard numerical methods and the software Xppaut to give the one-parameter and two-parameter bifurcation diagrams.

3.1. Predator–Prey Dynamics with Specialist Predator and Herd-Linear Functional Response: Boundedness, Equilibrium Points and Stability Analysis

We study the dynamics of the model by Rosenzweig and MacArthur [16] with herd-linear response $f(x) = ax^\alpha$, conversion factor e of captured prey into new predators, per capita natural mortality rate m for the predators and logistic growth $g(x) = rx(1 - \frac{x}{K})$ for the prey, where r denotes their net growth rate and K their carrying capacity,

$$\frac{dx}{dt} = rx\left(1 - \frac{x}{K}\right) - ax^\alpha y, \tag{6}$$

$$\frac{dy}{dt} = eax^\alpha y - my. \tag{7}$$

We show that the populations do not grow unbounded (we refer to the work by Bulai and Venturino in [7]). We define with $\psi(t) = x(t) + y(t)$ the total population density and, summing up the equations for the prey and predator populations, we obtain

$$\frac{d\psi(t)}{dt} = rx\left(1 - \frac{x}{K}\right) - (1-e)ax^\alpha y - m\psi(t) + mx. \tag{8}$$

We collect $\frac{d\psi(t)}{dt} + m\psi(t)$ on the left-hand side and drop the term $(1-e)ax^\alpha y > 0$ to obtain

$$\frac{d\psi(t)}{dt} + m\psi(t) \leq \max_x\left\{rx\left(1 - \frac{x}{K}\right) + mx\right\}. \tag{9}$$

The value of $\max_x\{rx(1 - \frac{x}{K}) + mx\}$ is at $x = \frac{(m+r)K}{2r}$ and, by substituting this, we obtain

$$\frac{d\psi(t)}{dt} + m\psi(t) \leq \frac{K(r+m)^2}{4r} \equiv \bar{M}. \tag{10}$$

We solve the equation for $\psi(t)$ and get the upper bound for $\psi(t)$, as well as $x(t)$ and $y(t)$

$$\psi(t) = e^{-mt}\left(\psi(0) - \frac{\bar{M}}{m}\right) + \frac{\bar{M}}{m} \leq \max\left\{\psi(0), \frac{\bar{M}}{m}\right\}. \tag{11}$$

To reduce the number of parameters, we introduce the dimensionless quantities $\tilde{x} = \frac{x}{K}$, $\tilde{y} = \frac{y}{K}$, $\tilde{t} = rt$, $\tilde{a} = \frac{aK^\alpha}{r}$, $\tilde{m} = \frac{m}{r}$. Applying the substitutions and dropping the tildes, we obtain the non-dimensional system

$$\frac{dx}{dt} = x(1-x) - ax^\alpha y, \tag{12}$$

$$\frac{dy}{dt} = eax^\alpha y - my, \tag{13}$$

with $g(x) = x(1-x)$ and $f(x) = ax^\alpha$.

The equilibria follow by setting the equations in (12) and (13) to zero. We obtain the trivial equilibrium points of the system

$$E_0 = (0,0), \quad E_1 = (1,0) \tag{14}$$

and the interior equilibrium $E^* = (x^*, y^*)$ with full expression below

$$E^* = \left(\left(\frac{m}{ae}\right)^{\frac{1}{\alpha}}, \frac{1}{a}\left(\frac{m}{ae}\right)^{\frac{1-\alpha}{\alpha}}\left[1 - \left(\frac{m}{ae}\right)^{\frac{1}{\alpha}}\right]\right). \tag{15}$$

Note that the interior equilibrium is positive if $\frac{m}{ae} < 1$.

We use the Jacobian matrix of the system in (12) and (13) to study the stability of the equilibria

$$J(x,y) = \begin{bmatrix} 1 - 2x - ax^{\alpha-1}\alpha y & -ax^\alpha \\ aex^{\alpha-1}\alpha y & aex^\alpha - m \end{bmatrix}. \tag{16}$$

The Jacobian evaluated at E_0 has possibly a singularity, but the instability of this point can be assessed looking back at the original Equations (12) and (13). With $y = 0$, and x near 0, the first equation behaves like $x' \approx rx$, so that x grows. Conversely, on $x = 0$ the second equation is $y' \approx -my$ and $y \to 0$. Hence, E_0 is a saddle. When evaluated at the equilibrium E_1, the determinant of the Jacobian matrix is $m - ae$ and is positive if $\frac{m}{ae} > 1$, that is, when the interior equilibrium is not feasible (i.e., does not exist or is negative). Under the same condition, the trace of the Jacobian at E_1, $ae - m - 1$, is negative and the equilibrium is stable.

For simplicity, we rewrite the Jacobian matrix evaluated at the interior equilibrium $E_* = (x^*, y^*)$ in terms of the functions $f(x)$ and $g(x)$

$$J(x^*, y^*) = \begin{bmatrix} g'(x^*) - f'(x^*)y^* & -f(x^*) \\ ef'(x^*)y^* & ef(x^*) - m \end{bmatrix} = \begin{bmatrix} f(x^*)\left(\frac{g(x^*)}{f(x^*)}\right)' & -\frac{m}{e} \\ ef'(x^*)y^* & 0 \end{bmatrix}. \quad (17)$$

The determinant of the matrix in (17) is $mf'(x^*)y^* > 0$ (since the functional response is an increasing function of the prey density), therefore the stability of the interior equilibrium depends on the sign of the trace $f(x^*)\left(\frac{g(x^*)}{f(x^*)}\right)'$, and, in particular, on the slope of the prey zero-growth isocline $\left(\frac{g(x^*)}{f(x^*)}\right)' = \frac{x^{-\alpha}[x(\alpha-2)+1-\alpha]}{a}\bigg|_{x=x^*}$ (see also Gause [17] and Gause et al. [18]). We obtain that the equilibrium E^* is stable if $\frac{m}{ae} > \left(\frac{\alpha-1}{\alpha-2}\right)^\alpha \in (0,1)$ (check Table 1 for a summary of the feasibility and stability conditions).

We conclude that a transcritical bifurcation occurs at $\frac{m}{ae} = 1$, where the interior equilibrium exchanges stability with the predator-free equilibrium. A Hopf bifurcation appears at $\frac{m}{ae} = \left(\frac{\alpha-1}{\alpha-2}\right)^\alpha$, as the eigenvalues of the community matrix become purely imaginary and the system converges to a stable limit cycle.

Table 1. Conditions for feasibility and stability of the equilibria of the system in (12) and (13). The trivial equilibrium E_0 is always unstable. TB: transcritical bifurcation. HB: Hopf bifurcation.

Condition	E_1	E_*	Bifurcation
$\frac{m}{ae} > 1$	asympt. stable	not feasible	
$\frac{m}{ae} = 1$			TB
$\left(\frac{\alpha-1}{\alpha-2}\right)^\alpha < \frac{m}{ae} < 1$	unstable	asympt. stable	
$\frac{m}{ae} = \left(\frac{\alpha-1}{\alpha-2}\right)^\alpha$			HB
$0 < \frac{m}{ae} < \left(\frac{\alpha-1}{\alpha-2}\right)^\alpha$	unstable	unstable	

3.2. Predator–Prey Dynamics with Generalist Predator and Herd-Linear Functional Response: Boundedness, Equilibrium Points and Stability Analysis

We study the dynamics of the model by Rosenzweig and MacArthur [16] with herd-linear response $f(x) = ax^\alpha$, conversion factor e of captured prey into new predators and per capita natural mortality rate m for the predators. We assume logistic growth for both the prey and the predator species, $g_x(x) = rx\left(1 - \frac{x}{K_x}\right)$ and $g_y(x) = sx\left(1 - \frac{x}{K_y}\right)$, respectively, where r is the net growth rate of the prey and K_x their carrying capacity, while s denotes the predators' reproduction rate, i.e., not discounted by deaths,

$$\frac{dx}{dt} = rx\left(1 - \frac{x}{K_x}\right) - ax^\alpha y, \quad (18)$$

$$\frac{dy}{dt} = sy\left(1 - \frac{y}{K_y}\right) + eax^\alpha y - my. \quad (19)$$

In this way, the predators are subjected to intraspecific competition, which occurs at rate $\frac{s}{K_y}$.

To check that the populations do not grow unbounded, we set $\psi(t) = x(t) + y(t)$ and, by repeating the steps in Section 3.1, we get the differential equation for the total population

$$\frac{d\psi(t)}{dt} + m\psi(t) \leq \max_{x,y}\left\{rx\left(1 - \frac{x}{K_x}\right) + sy\left(1 - \frac{y}{K_y}\right) + mx\right\}. \tag{20}$$

We differentiate the right-hand term with respect to x and y to get the local maximum $\left(\frac{(r+m)}{2r}K_x, \frac{K_y}{2}\right)$. By substitution in the equation above, we obtain

$$\frac{d\psi(t)}{dt} + m\psi(t) \leq \frac{(r+m)^2}{4r}K_x + \frac{s}{4}K_y \equiv \bar{M}. \tag{21}$$

Therefore, the solution for the total population reads

$$\psi(t) = e^{-mt}\left(\psi(0) - \frac{\bar{M}}{m}\right) + \frac{\bar{M}}{m} \leq \max\left\{\psi(0), \frac{\bar{M}}{m}\right\}, \tag{22}$$

where the upper bound is applicable also for $x(t)$ and $y(t)$.

To obtain the non-dimensional version of the system in (18) and (19), we consider the dimensionless variables and parameters $\tilde{x} = \frac{x}{K_x}, \tilde{y} = \frac{y}{K_y}, \tilde{t} = rt, \tilde{a} = \frac{K_y}{K_x^{1-\alpha}r}a, \tilde{s} = \frac{s}{r}, \tilde{e} = \frac{K_x}{K_y}e, \tilde{m} = \frac{m}{r}$. We drop the tildes and obtain the dimensionless system

$$\frac{dx}{dt} = x(1-x) - ax^\alpha y, \tag{23}$$

$$\frac{dy}{dt} = sy(1-y) + eax^\alpha y - my, \tag{24}$$

with $g_x(x) = x(1-x)$, $g_y(y) = sy(1-y)$ and $f(x) = ax^\alpha$.

We proceed with computing the equilibria. The trivial equilibria are

$$E_0 = (0,0), \quad E_1 = (1,0), \quad E_2 = \left(0, 1 - \frac{m}{s}\right), \tag{25}$$

with E_2 feasible if $m < s$. The interior equilibria are given by the intersection of the isoclines

$$y = \frac{(1-x)x^{1-\alpha}}{a} \tag{26}$$

and

$$y = 1 - \frac{m}{s} + \frac{ea}{s}x^\alpha. \tag{27}$$

Note that the isocline in (26) intersects the x-axis at $(0,0)$ and $(1,0)$ and has a maximum at $x = \frac{1-\alpha}{2-\alpha} < \frac{1}{2}$ for $0 < \alpha < 1$, while the isocline in (27) intersects the vertical axis at $\left(0, 1 - \frac{m}{s}\right)$ and is a root function translated by $1 - \frac{m}{s}$ and dilated by $\frac{ea}{s}$. Therefore, if the intersection point of the isocline in (27) lies in the positive quadrant, i.e., if $m < s$, we find three different configurations for the phase plane: the two isoclines can intersect at most twice at E_{*1} and E_{*2} if $1 - \frac{m}{s} < \frac{1}{a(2-\alpha)}\left(\frac{1-\alpha}{2-\alpha}\right)^{1-\alpha} - \frac{ea}{s}\left(\frac{1-\alpha}{2-\alpha}\right)^\alpha$, or be tangent at $E_* = \left(\frac{1-\alpha}{2-\alpha}, 1 - \frac{m}{s} + \frac{ea}{s}\left(\frac{1-\alpha}{2-\alpha}\right)^\alpha\right)$ when $1 - \frac{m}{s} = \frac{1}{a(2-\alpha)}\left(\frac{1-\alpha}{2-\alpha}\right)^{1-\alpha} - \frac{ea}{s}\left(\frac{1-\alpha}{2-\alpha}\right)^\alpha$ or never intersect in $x \in (0,1)$ when $1 - \frac{m}{s} > \frac{1}{a(2-\alpha)}\left(\frac{1-\alpha}{2-\alpha}\right)^{1-\alpha} - \frac{ea}{s}\left(\frac{1-\alpha}{2-\alpha}\right)^\alpha$. The equilibria are obtained as the positive roots of the curve

$$\phi(x) = 1 - \frac{m}{s} + \frac{ea}{s}x^\alpha - (1-x)\frac{x^{1-\alpha}}{a} \tag{28}$$

and the non-negativity of y is ensured by the condition

$$\left(\frac{m-1}{ea}\right)^{\frac{1}{\alpha}} < x < 1. \tag{29}$$

When $m > s$, the isocline in (27) intersects the vertical axes at $y = 1 - \frac{m}{s} < 0$ and we find at most one interior equilibrium E_*. We obtain the feasibility condition for E_* by imposing that the curve in (27) takes positive values at $x = 1$, that is, if $m < s + ea$.

The Jacobian matrix of the system in (23) and (24) is given by

$$J(x,y) = \begin{bmatrix} 1 - 2x - ax^{\alpha-1}\alpha y & -ax^\alpha \\ aex^{\alpha-1}\alpha y & s - 2sy + aex^\alpha - m \end{bmatrix}. \tag{30}$$

The equilibrium E_0, restricting the analysis to the trajectories on the coordinate axes, is seen to be either an unstable node if $m < s$, or a saddle if $m > s$. The prey-only equilibrium E_1 is a stable node if $m > s + ae$, otherwise the steady state is an unstable saddle. Under its feasibility condition $m < s$, the equilibrium E_2 is always an unstable saddle. We summarise the feasibility and stability conditions studied above in Tables 2 and 3.

We rewrite the Jacobian matrix evaluated at the interior equilibrium in terms of functions $f(x)$, $g_x(x)$ and $g_y(y)$:

$$J(x^*, y^*) = \begin{bmatrix} g'_x(x^*) - f'(x^*)y^* & -f(x^*) \\ ef'(x^*)y^* & g'_y(y) + ef(x^*) - m \end{bmatrix} = \begin{bmatrix} f(x^*)\left(\frac{g_x(x^*)}{f(x^*)}\right)' & -f(x^*) \\ ef'(x^*)y^* & -y^* \end{bmatrix}. \tag{31}$$

The trace and the determinant at the interior equilibrium are given by

$$\text{tr} J(x^*, y^*) = f(x^*)\left(\frac{g_x(x^*)}{f(x^*)}\right)' - y^*, \tag{32}$$

$$\det J(x^*, y^*) = -y^* f(x^*)\left(\frac{g_x(x^*)}{f(x^*)}\right)' + ef(x^*)f'(x^*)y^*. \tag{33}$$

When only one interior equilibrium exists and is positive, the signs of the trace and the determinant determine its asymptotic stability, more specifically if $\text{tr} J(x^*, y^*) < 0$ and $\det J(x^*, y^*) > 0$.

It seems rather difficult to obtain more analytical details on the stability of the equilibria and bifurcations for the model in (23) and (24). If possible, a more detailed analysis will be the topic of a future work.

Table 2. Conditions for the feasibility and stability of the equilibria of the system in (23) and (24).

Condition	E_0	E_1	E_2	E_{*1}	E_{*2}
$m < s$	unstable	unstable	unstable	See Table 3	See Table 3
$s < m < s + ae$	unstable	unstable	not feas.	stable	not feasible
$m > s + ae$	unstable	asympt. stable	not feasible	not feasible	not feasible

Table 3. Conditions for feasibility of the interior equilibria of the system in (23) and (24) when $m < s$.

Condition	E_{*1}	E_{*2}
$1 - \frac{m}{s} < \frac{1}{a(2-\alpha)}\left(\frac{1-\alpha}{2-\alpha}\right)^{1-\alpha} - \frac{ea}{s}\left(\frac{1-\alpha}{2-\alpha}\right)^\alpha$	feasible	feasible
$1 - \frac{m}{s} = \frac{1}{a(2-\alpha)}\left(\frac{1-\alpha}{2-\alpha}\right)^{1-\alpha} - \frac{ea}{s}\left(\frac{1-\alpha}{2-\alpha}\right)^\alpha$	feasible	not feasible
$1 - \frac{m}{s} > \frac{1}{a(2-\alpha)}\left(\frac{1-\alpha}{2-\alpha}\right)^{1-\alpha} - \frac{ea}{s}\left(\frac{1-\alpha}{2-\alpha}\right)^\alpha$	not feasible	not feasible

3.3. Predator–Prey Dynamics with Specialist Predator and Herd-Holling Type II Functional Response: Boundedness, Equilibrium Points and Stability Analysis

We study the dynamics of the model by Rosenzweig and MacArthur [16] with the herd-Holling type II functional response $f(x) = \frac{ax^\alpha}{1+ahx^\alpha}$ derived in Section 2, conversion factor e of captured prey into new predators and per capita natural mortality rate m for the predators, logistic growth $g(x) = rx(1 - \frac{x}{K})$ for the prey, where r denotes their net growth rate and K their carrying capacity,

$$\frac{dx}{dt} = rx\left(1 - \frac{x}{K}\right) - \frac{ax^\alpha}{1 + ahx^\alpha}y, \tag{34}$$

$$\frac{dy}{dt} = e\frac{ax^\alpha}{1 + ahx^\alpha}y - my. \tag{35}$$

The total population $\psi(t) = x(t) + y(t)$ verifies

$$\psi(t) \leq \max\left\{\psi(0), \frac{\bar{M}}{m}\right\}, \tag{36}$$

with $\bar{M} = \frac{K(r+m)^2}{4r}$ as in Section 3.1.

We obtain the dimensionless version of the model by applying the substitutions $\tilde{x} = \frac{x}{K}$, $\tilde{y} = \frac{y}{K}$, $\tilde{t} = rt$, $\tilde{a} = \frac{aK^\alpha}{r}$, $\tilde{h} = rh$, $\tilde{m} = \frac{m}{r}$. We drop the tildes and get the equations

$$\frac{dx}{dt} = x(1 - x) - \frac{ax^\alpha}{1 + ahx^\alpha}y, \tag{37}$$

$$\frac{dy}{dt} = e\frac{ax^\alpha}{1 + ahx^\alpha}y - my, \tag{38}$$

with $g(x) = x(1 - x)$ and $f(x) = \frac{ax^\alpha}{1+ahx^\alpha}$.

We compute the equilibria by setting the equations in (37) and (38) to zero. The trivial equilibria are $E_0 = (0,0)$ and $E_1 = (1,0)$, while the unique interior equilibrium $E_* = (x^*, y^*)$ has full expression

$$E_* = \left(\left(\frac{m}{ea - mah}\right)^{\frac{1}{\alpha}}, \frac{e}{m}x^*(1 - x^*)\right) \tag{39}$$

and exists and is positive if and only if $mah + m < ea$.

The Jacobian matrix corresponding to the system in (37) and (38) is given by

$$J(x,y) = \begin{bmatrix} 1 - 2x - \frac{a\alpha x^{\alpha-1}}{(1+ahx^\alpha)^2}y & \frac{ax^\alpha}{1+ahx^\alpha} \\ e\frac{a\alpha x^{\alpha-1}}{(1+ahx^\alpha)^2}y & e\frac{ax^\alpha}{1+ahx^\alpha} - m \end{bmatrix}. \tag{40}$$

The origin is unstable, being a saddle, a fact that is shown restricting the system to the coordinate axes, as previously done for the system (12) and (13). The equilibrium E_1 is stable if and only if $ea < mah + m$ (under this condition the determinant of the Jacobian matrix at the equilibrium is positive and the trace is negative). The prey-only equilibrium changes its stability at $ea = mah + m$ when a transcritical bifurcation occurs. We can use the same formulation as in (17) for the Jacobian evaluated at the interior equilibrium, which for convenience we reproduce here

$$J(x^*, y^*) = \begin{bmatrix} g'(x^*) - f'(x^*)y^* & -f(x^*) \\ ef'(x^*)y^* & ef(x^*) - m \end{bmatrix} = \begin{bmatrix} f(x^*)\left(\frac{g(x^*)}{f(x^*)}\right)' & -\frac{m}{e} \\ ef'(x^*)y^* & 0 \end{bmatrix}. \tag{41}$$

As the determinant of the Jacobian matrix is always positive, the stability of the interior equilibrium depends on the sign of the trace, in particular on the slope of the predator isocline, $\left(\frac{g(x^*)}{f(x^*)}\right)' = \frac{x^{-\alpha}[x(\alpha-2)+1-\alpha+ahx^\alpha(1-2x)]}{a}\bigg|_{x=x^*}$. For the same reason as in Section 3.1, the system in (37) and (38) undergoes a Hopf bifurcation when $\left(\frac{g(x^*)}{f(x^*)}\right)' = 0$ and converges to a stable limit cycle for $\left(\frac{g(x^*)}{f(x^*)}\right)' > 0$, otherwise it converges to the interior equilibrium E_*. In Table 4 we give a summary of the feasibility and stability conditions of the equilibria.

Table 4. Conditions for feasibility and stability of the equilibria of the system in (37) and (38). TB: transcritical bifurcation. HB: Hopf bifurcation.

Condition	E_0	E_1	E_*	Bifurcation
$ea < mah + m$	unstable	asympt. stable	not feasible	
$ea = mah + m$				TB
$ea > mah + m$, $\left(\frac{g(x^*)}{f(x^*)}\right)' < 0$	unstable	unstable	asympt. stable	
$ea > mah + m$, $\left(\frac{g(x^*)}{f(x^*)}\right)' = 0$				HB
$ea > mah + m$, $\left(\frac{g(x^*)}{f(x^*)}\right)' > 0$	unstable	unstable	unstable	

3.4. Predator–Prey Dynamics with Generalist Predator and Herd-Holling Type II Functional Response: Boundedness, Equilibrium Points and Stability Analysis

We study the dynamics of the model by Rosenzweig and MacArthur [16] with the herd-Holling type II functional response $f(x) = \frac{ax^\alpha}{1+ahx^\alpha}$ derived in Section 2, conversion factor e of captured prey into new predators and per capita natural mortality rate m for the predators. We assume logistic growth for both the prey and the predator species, $g_x(x) = rx\left(1 - \frac{x}{K_x}\right)$ and $g_y(x) = sx\left(1 - \frac{x}{K_y}\right)$, respectively, where r is the net growth rate of the prey and K_x their carrying capacity, while s denotes the predators' reproduction rate,

$$\frac{dx}{dt} = rx\left(1 - \frac{x}{K_x}\right) - \frac{ax^\alpha}{1+ahx^\alpha}y, \tag{42}$$

$$\frac{dy}{dt} = sy\left(1 - \frac{y}{K_y}\right) + e\frac{ax^\alpha}{1+ahx^\alpha}y - my. \tag{43}$$

Once again, note the second term in the predators' equation, whose coefficient $\frac{s}{K_y}$ models predators intraspecific competition.

The boundedness of the populations is ensured by the condition on the total population density $\psi(t) = x(t) + y(t)$

$$\psi(t) \leq \max\left\{\psi(0), \frac{\bar{M}}{m}\right\}, \tag{44}$$

with $\bar{M} = \frac{(r+m)^2}{4r}K_x + \frac{s}{4}K_y$ as in Section 3.1.

We use the dimensionless quantities $\tilde{x} = \frac{x}{K_x}$, $\tilde{y} = \frac{y}{K_y}$, $\tilde{t} = rt$, $\tilde{a} = \frac{K_y}{K_x^{1-\alpha}r}a$, $\tilde{h} = r\frac{K_x}{K_y}h$, $\tilde{s} = \frac{s}{r}$, $\tilde{e} = \frac{K_x}{K_y}e$, $\tilde{m} = \frac{m}{r}$ to obtain the dimensionless system of equations

$$\frac{dx}{dt} = x(1-x) - \frac{ax^\alpha}{1+ahx^\alpha}y, \tag{45}$$

$$\frac{dy}{dt} = sy(1-y) + e\frac{ax^\alpha}{1+ahx^\alpha}y - my, \tag{46}$$

with $g_x(x) = x(1-x)$, $g_y(y) = sy(1-y)$ and $f(x) = \frac{ax^\alpha}{1+ahx^\alpha}$.

The corresponding trivial equilibria correspond to the ones in Section 3.2 and are given by

$$E_0 = (0,0), \quad E_1 = (1,0), \quad E_2 = \left(0, 1 - \frac{m}{s}\right), \tag{47}$$

with E_2 being feasible if $m < s$. The interior equilibria are given by the intersection of the isoclines

$$y = \frac{x(1-x)(1+ahx^\alpha)}{ax^\alpha} \tag{48}$$

and
$$y = 1 - \frac{m}{s} + \frac{eax^\alpha}{s(1+ahx^\alpha)}. \tag{49}$$

The isocline in (48) intersects the horizontal axis at $(0,0)$ and $(1,0)$, while the isocline in (49) intersects the vertical axis at $(0, 1 - \frac{m}{s})$. As for the model with generalist predator and linear functional response in Section 3.2, we may expect that, under some conditions, the system admits two interior equilibria. However, given the formulation of the isoclines in (48) and (49), it seems difficult to find explicit analytical results and we refer to the next Section 3.5 for more details on the interior equilibria and their feasibility and stability conditions.

We obtain the Jacobian matrix to check the stability of the trivial equilibria:

$$J(x,y) = \begin{bmatrix} 1 - 2x - \frac{a\alpha x^{\alpha-1}}{(1+ahx^\alpha)^2}y & \frac{ax^\alpha}{1+ahx^\alpha} \\ e\frac{a\alpha x^{\alpha-1}}{(1+ahx^\alpha)^2}y & s - 2sy + e\frac{ax^\alpha}{1+ahx^\alpha} - m \end{bmatrix}. \tag{50}$$

Again, restricting the analysis to the trajectories on the coordinate axes, we obtain that E_0 is an unstable node for $m < s$, or a saddle if $m > s$. We find that the origin E_0 is an unstable saddle, as well as the predator-only equilibrium E_2. The prey-only equilibrium E_1 is a stable node if $m > s + \frac{ea}{1+ah}$, an unstable saddle otherwise. We give a summary of these results in Table 5.

Table 5. Conditions for feasibility and stability of the trivial equilibria of the system in (45) and (46).

Condition	E_0	E_1	E_2
$m < s + \frac{ea}{1+ah}$	unstable	unstable	unstable
$m > s + \frac{ea}{1+ah}$, $\left(\frac{g(x^*)}{f(x^*)}\right)' < 0$	unstable	asympt. stable	unstable

3.5. One-Parameter and Two-Parameter Bifurcation Diagrams

In this section, we proceed with the bifurcation analysis. We give the one-parameter bifurcation diagrams and vary either the value of the predator mortality rate, m, or the herding index, α, when possible. Additionally, we obtain the two-parameter bifurcation diagrams with respect to the parameter pairs $(m, \star)$ or $(\alpha, \star)$, where $\star$ equals one of the other parameters in the model. Note that in the numerical simulations we use the model and parameters prior to non-dimensionalisation, to obtain a complete analysis with respect to all the model parameters.

We first study the predator–prey model with specialist predator and herd-linear functional response. In the one-parameter bifurcation plots, we fix the parameter values of the model in (6) and (7) as in Table 6 and we call it the nominal set (hypothetical values).

Table 6. Nominal set of parameter values for the model in (6) and (7).

Parameter	Description	Value
r	prey net reproduction rate	1
K	prey carrying capacity	5
a	predation rate	1
α	herd exponent	0.7
e	conversion factor	0.5
m	natural mortality rate (predators)	1

In Figure 1 we give the one-parameter bifurcation diagrams with respect to the natural mortality rate of the predators, m. Note that when $m = 0.5526$, the system undergoes a supercritical Hopf bifurcation (HB) and a stable limit cycle appears. At $m = 1.543$, a transcritical bifurcation occurs, where the coexistence equilibrium loses its stability and the predator-free equilibrium becomes stable.

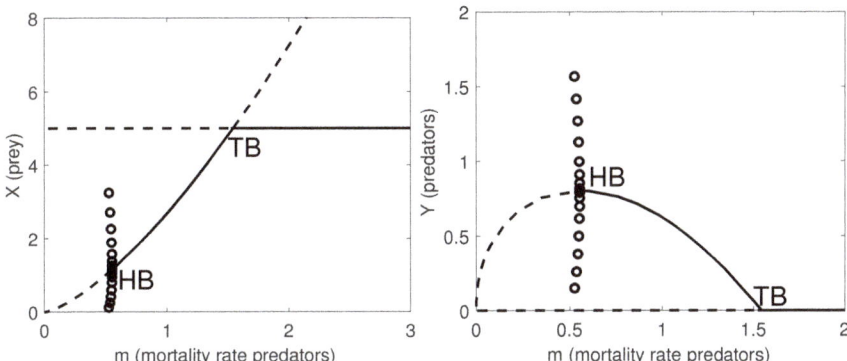

Figure 1. One-parameter bifurcation diagram with respect to m. **Left panel**: the prey population dynamics. **Right panel**: the predator population dynamics. Thick lines: stable equilibria; dashed lines: unstable equilibria; HB: supercritical Hopf bifurcation point where a stable limit cycle arises with maximum amplitude given by the amplitude of the HB (circles); TB: transcritical bifurcation point where the coexistence equilibrium exchanges its stability with the prey-only equilibrium. The remaining parameter values are as in Table 6.

To complete the analysis, in Figure 2 we have plotted all the possible two-parameter bifurcation diagrams for $(m, \star)$, with $\star = a, K, a, \alpha$ or e. Note that the HB curve appears in all two-parameter bifurcation diagrams, but only in the plots where we vary (m, K), (m, a) and (m, e) it occurs for every value of m. When we let r vary, the HB curve is present only at $m = 0.5526$ (Figure 2, top left); similarly, when we allow α to change, the HB occurs only for $m \leq 0.5$ (Figure 2, bottom left).

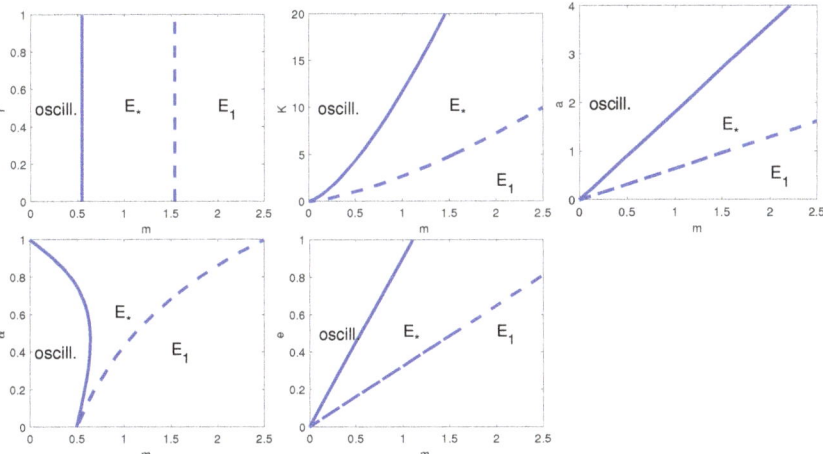

Figure 2. Two-parameter bifurcation diagrams with respect to m and, from **top-left** to **bottom-right**, r, K, a, α and e. Thick line: HB curve; dashed line: transcritical bifurcation curve. E_* is the coexistence equilibrium and E_1 the prey-only equilibrium. The remaining parameter values are as in Table 6.

As a second example, we analyse the predator–prey dynamics with generalist predator and herd-linear functional response. In Table 7 we give the nominal set of parameter values for the model in (18) and (19).

Table 7. Nominal set of parameter values for the model in (18) and (19).

Parameter	Description	Value
r	prey net reproduction rate	0.5
K_x	prey carrying capacity	5
a	predation rate	1
α	herd exponent	0.7
s	predator reproduction rate	0.5
K_y	predator carrying capacity	5
e	conversion factor	0.5
m	natural mortality rate (predators)	1

In Figure 3 we give the one-parameter bifurcation diagram with respect to the herd exponent, α. Here a supercritical HB occurs at $\alpha = 0.5995$ and a subcritical HB appears at $\alpha = 0.1476$.

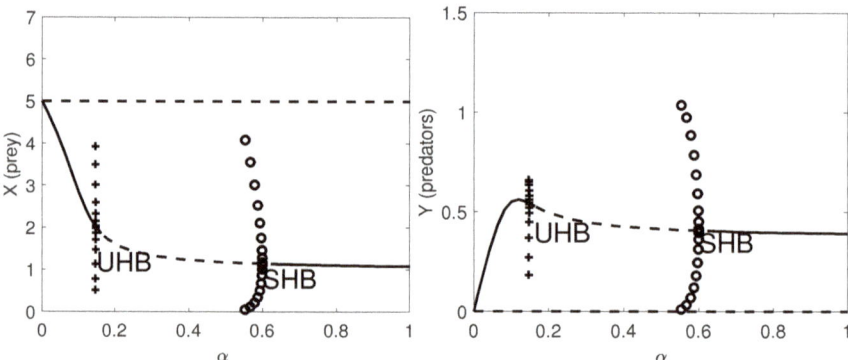

Figure 3. One-parameter bifurcation diagram with respect to α. **Left panel**: the prey population dynamics. **Right panel**: the predator population dynamics. Thick lines: stable equilibria; dashed lines: unstable equilibria; SHB: supercritical Hopf bifurcation point where a stable limit cycle arises with maximum amplitude given by the amplitude of the SHB (circles); UHB: subcritical Hopf bifurcation point, where an unstable limit cycle arises with maximum amplitude given by the amplitude of the UHB (pluses); TB: transcritical bifurcation point where the coexistence equilibrium exchanges its stability with the prey-only equilibrium. The remaining parameter values are as in Table 7.

The two-parameters bifurcation diagrams with respect to $(\alpha, \star)$, with $\star = r, K_x, a, s, K_y, e,$ and m for the model with Equations (18) and (19) are given in Figure 4. When we vary the parameter pair (α, a), a HB appears for all values of α independently of the value of a, while for the remaining cases the HB is present only for some parameter values. Moreover, we observe that only when we vary the parameter pair (α, m), the transcritical bifurcation curve is present. Finally, if one of the parameters $\star = K_x, a, s, K_y,$ and e are below certain threshold values, with the other parameter values fixed as in Figure 4, the coexistence equilibrium is asymptotically stable; analogously, when r is above a certain threshold value the system converges to the interior equilibrium.

In this paragraph, we focus on the predator–prey dynamics with specialist predator and herd-Holling type II functional response. In Table 8 we list the parameter values for the model in (34) and (35).

Table 8. Nominal set of parameter values for the model in (34) and (35).

Parameter	Description	Value
r	prey net reproduction rate	0.5
K	prey carrying capacity	5
a	predation rate	1
α	herd exponent	0.7
h	handling time	0.2
e	conversion factor	0.5
m	natural mortality rate (predators)	1

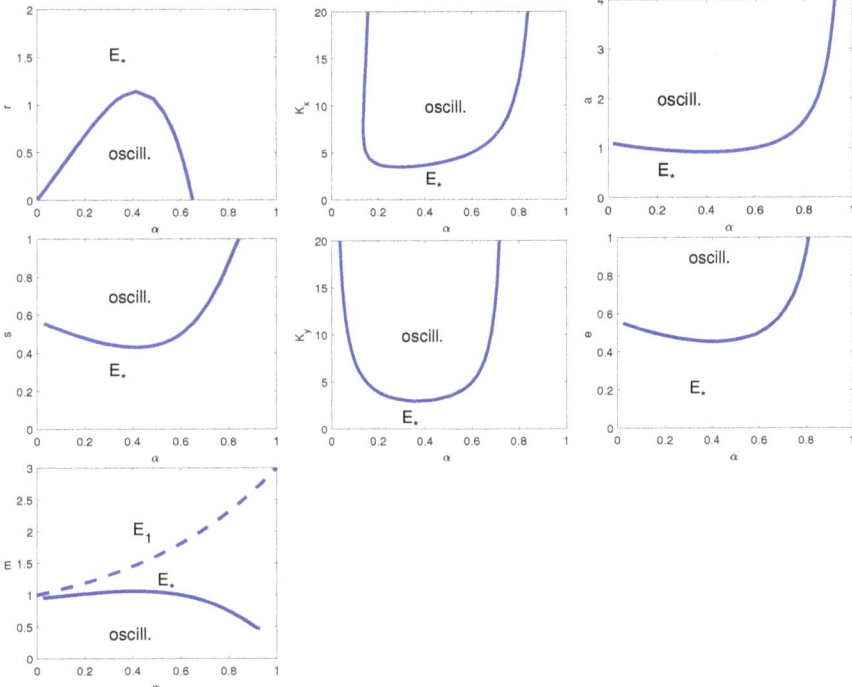

Figure 4. Two-parameter bifurcation diagram with respect to α, and, from top left to bottom right, r, K_x, a, s, K_y, e and m. Thick line: HB curve; dashed line: transcritical bifurcation curve. E_* is the coexistence equilibrium and E_1 the prey-only equilibrium. The remaining parameter values are as in Table 7.

In Figure 5 we obtain qualitatively similar results as for (6) and (7) for the one-parameter bifurcation diagrams with respect to the natural mortality rate of the predators, m.

The dynamic described in Figure 6 is similar to the one in Figure 2. There are two main differences: in the two-parameter bifurcation diagrams with respect to (m, K) and (m, a) the HB curve is more concave; when we vary the parameter pair (m, h) (this case is not present in Figure 2), one can see that the HB occurs for all values of h and for m smaller than a threshold value.

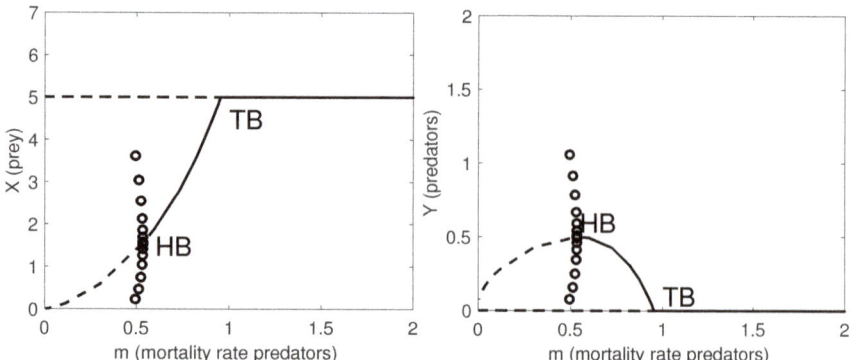

Figure 5. One-parameter bifurcation diagram with respect to the natural mortality rate of the predators, m. **Left panel**: the prey population dynamics; **Right panel**: the predator population dynamics. Thick lines: stable equilibria; dashed lines: unstable equilibria. HB: supercritical Hopf bifurcation point where a stable limit cycle arises with maximum amplitude given by the amplitude of the HB (circles); TB: transcritical bifurcation point where the coexistence equilibrium exchanges its stability with the prey-only equilibrium. The remaining parameter values are as in Table 8.

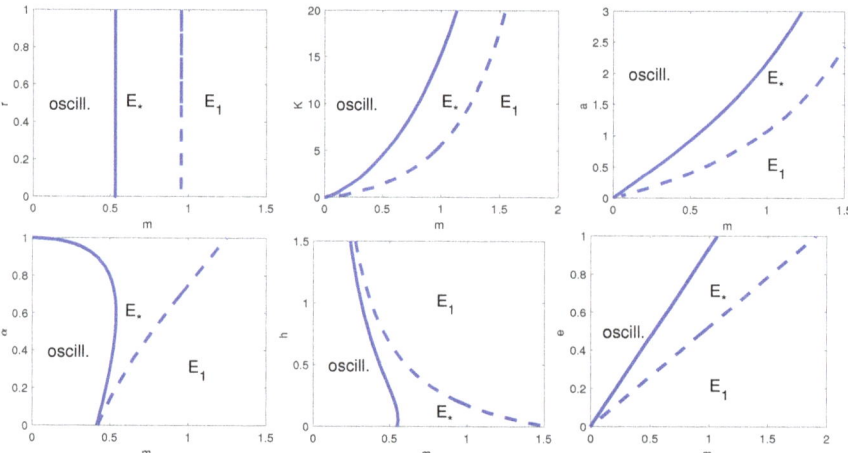

Figure 6. Two-parameter bifurcation diagram with respect to m and, from top left to bottom right, r, K, a, α, h and e. Thick line: HB curve; dashed line: transcritical bifurcation curve. E_* is the coexistence equilibrium and E_1 the prey-only equilibrium. The remaining parameter values are as in Table 8.

Finally, we study the predator–prey dynamics with generalist predator and herd-Holling type II functional response. In Table 9 we list the nominal set of parameter values for model in (42) and (43). This is the most general model that we study which encompasses all the previously considered cases.

Both the one-parameter bifurcation diagram with respect to m, and two-parameter bifurcation diagrams with respect to $(m, \star)$, with $\star = r, K_x, a, \alpha, h, s, K_y$ and e show behaviours similar to the previous models, see Figures 7 and 8, respectively. It is worth noting that the results for models (34) and (35) are different as we give the two-parameter bifurcation diagrams with respect to the herd exponent as first parameter, while we refer to the predator natural mortality rate in the other two-parameter bifurcation plots.

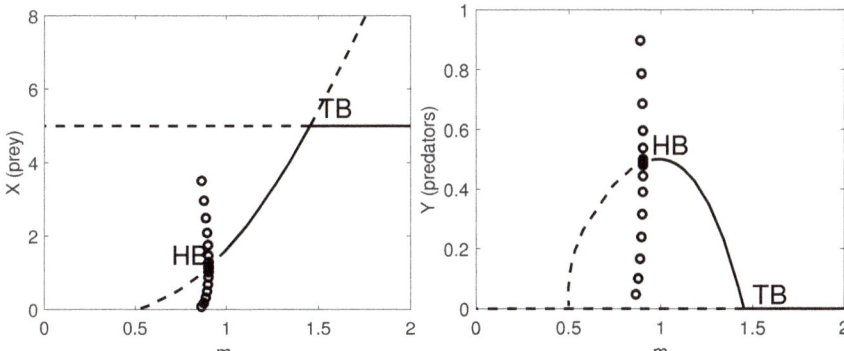

Figure 7. One-parameter bifurcation diagram with respect to m. **Left panel**: the prey population dynamics. **Right panel**: the predator population dynamics. Thick lines: stable equilibria; dashed lines: unstable equilibria; HB: supercritical Hopf bifurcation point where a stable limit cycle arises with maximum amplitude given by the amplitude of the HB (circles); TB: transcritical bifurcation point where the coexistence equilibrium exchanges its stability with the prey-only equilibrium. The remaining parameter values are as in Table 9.

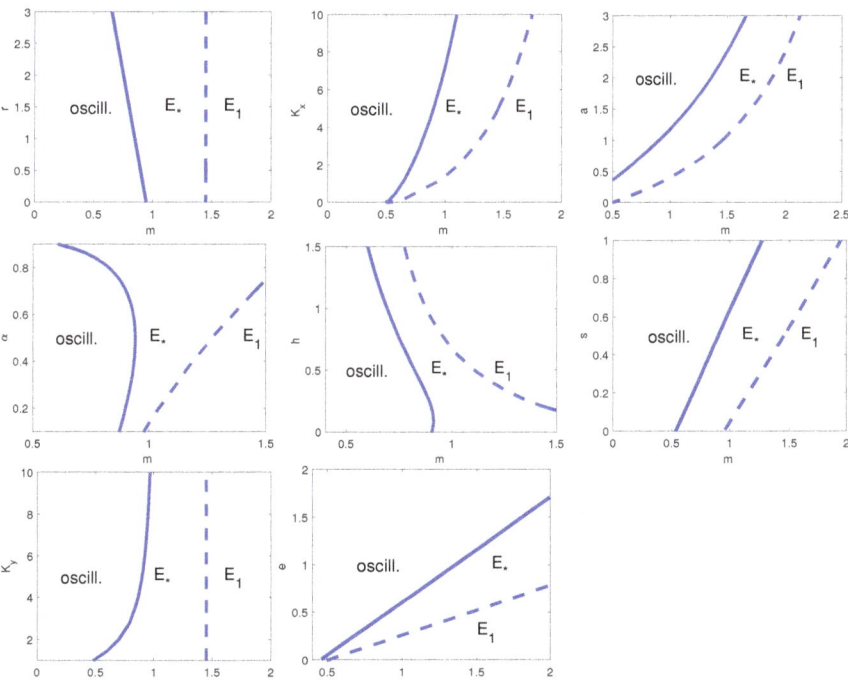

Figure 8. Two-parameters bifurcation diagram with respect to m and, from top left to bottom right, r, K_x, a, α, h, s, K_y, and e. Thick line: HB curve; dashed line: transcritical bifurcation curve. E_* is the coexistence equilibrium and E_1 the prey-only equilibrium. The remaining parameter values are as in Table 9.

Table 9. Nominal set of parameter values for the third model (42) and (43).

Parameter	Description	Value
r	prey net reproduction rate	0.5
K_x	prey carrying capacity	5
a	predation rate	1
α	herd exponent	0.7
h	handling time	0.2
s	predator reproduction rate	0.5
K_y	predator carrying capacity	5
e	conversion factor	0.5
m	natural mortality rate (predators)	1

4. Conclusions

The aim of this paper is to formalise, by means of illustrative examples, how prey herd behaviour can be modelled with ordinary differential equations from first principles. We propose four different Rosenzweig–MacArthur predator–prey models where the prey gather in herds and only the individuals on the edge are subjected to the predators' attacks.

We use the mechanistic approach and time-scale separation method to derive from the individual mechanisms a Holling type II–like functional response for the predators, the *herd-Holling type II functional response*, which takes into account the prey herd shape. As strongly emphasised in [11–15], the strength of the mechanistic approach lies in the possibility of interpreting the population equations in terms of individual state transition parameter rates.

Moreover, by introducing the so-called *herding index* α (see [5–10]), we model the density of the prey x on the edge of the herd as x^α and we are able to track information on the herd shape in a simple system of ordinary differential equations for the population dynamics.

We focus our attention on the ecological dynamics with one prey and one predator species, including specialist versus generalist predators, i.e., no predators' intraspecific competition versus predators' intraspecific competition. Such antagonistic behaviour has been widely observed in population ecology, especially in aquatic species and insects, and has been proven to deeply affect *niche* expansion and speciation (see, for example, [19–23]). We further assume either herd-linear or herd-Holling type II functional responses for the predators.

Unlike the two simple cases with specialist predator where both the analytical and numerical results on the attractors and bifurcations for the population dynamics given in this paper are exhaustive, for the two models assuming intraspecific competition there is room for a more detailed study. Indeed, we have found that such ecological dynamics can give rise to multiple stable attractors (cycles and steady states), while our current analysis comprises only the cases with one interior equilibrium.

Moreover, given the evidence of possible chaotic behaviour in models with prey herd behaviour, as found for instance in the recent paper [24], the analysis can be widened in both analytical and numerical directions, by introducing an adequate change of variable. This has the advantage of simplifying the exponential term in the population equations as in [7] and analysing two-bifurcation diagrams in the same way as performed in [25].

Further extensions of our models may include more complex types of predators and prey-predator dynamics, e.g., generalist predators with respect to prey switching and prey–predator–superpredator models, where the superpredator species feeds on the prey and alternative food sources, such as cannibalising the weaker and younger predator individuals (see, for example, the state transitions and time-scale separation in [26,27] for a system with one predator species with cannibalistic tendencies). However, the greater the number of states and state transitions, the more complicated the population dynamics. Extending the mechanistic derivation of the population equations from individual-level interactions to ecosystems with multiple prey and predator species can further increase the complexity of the model, which may become parameter-heavy and will require advanced numerical methods.

Author Contributions: Formal analysis, C.B.; investigation, C.B. and I.M.B.; methodology, C.B., I.M.B. and E.V.; software, I.M.B.; writing—review and editing, C.B., I.M.B. and E.V. All authors have read and agreed to the published version of the manuscript.

Funding: C. Berardo was supported by the Academy of Finland, Centre of Excellence in Analysis and Dynamics Research. I. M. Bulai was supported by MIUR through PON-AIM Linea 1 (AIM1852570-1).

Institutional Review Board Statement: Not applicable.

Informed Consent Statement: Not applicable.

Data Availability Statement: Not applicable.

Acknowledgments: This research has been accomplished within the UMI Group TAA "Approximation Theory and Application".

Conflicts of Interest: The authors declare no conflict of interest.

References

1. Laurie, H.; Venturino, E.; Bulai, I.M. Herding Induced by Encounter Rate, with Predator Pressure Influencing Prey Response. *Curr. Trends Dyn. Syst. Biol. Nat. Sci.* **2020**, *21*, 63.
2. Liu, W.M.; Hethcote, H.W.; Levin, S.A. Dynamical behavior of epidemiological models with nonlinear incidence rates. *J. Math. Biol.* **1987**, *25*, 359–380. [CrossRef]
3. Bhattacharyya, R.; Mukhopadhyay, B. On an epidemiological model with nonlinear infection incidence: Local and global perspective. *Appl. Math. Model.* **2011**, *35*, 3166–3174. [CrossRef]
4. Materassi, M. Some fractal thoughts about the COVID-19 infection outbreak. *Chaos Solitons Fract. X* **2019**, *4*, 100032. [CrossRef]
5. Ajraldi, V.; Pittavino, M.; Venturino, E. Modelling herd behavior in population systems. *Nonlinear Anal. Real World Appl.* **2011**, *12*, 2319–2338. [CrossRef]
6. Gimmelli, G.; Kooi, B.W.; Venturino, E. Ecoepidemic models with prey group defense and feeding saturation. *Ecol. Complex.* **2015**, *22*, 50–58. [CrossRef]
7. Bulai, I.M.; Venturino, E. Shape effects on herd behavior in ecological interacting population models. *Math. Comput. Simul.* **2017**, *141*, 40–55. [CrossRef]
8. Djilali, S. Impact of prey herd shape on the predator-prey interaction. *Chaos Solitons Fract.* **2019**, *120*, 139–148. [CrossRef]
9. Bentout, S.; Djilali, S.; Kumar, S. Mathematical analysis of the influence of prey escaping from prey herd on three species fractional predator-prey interaction model. *Phys. A Stat. Mech. Its Appl.* **2021**, *572*, 125840. [CrossRef]
10. Djilali, S.; Ghanbari, B. Dynamical behavior of two predators–one prey model with generalized functional response and time-fractional derivative. *Adv. Differ. Equ.* **2021**, *2021*, 235. [CrossRef]
11. Berardo, C.; Geritz, S.; Gyllenberg, M.; Raoul, G. Interactions between different predator–prey states: A method for the derivation of the functional and numerical response. *J. Math. Biol.* **2020**, *80*, 2431–2468. [CrossRef]
12. Geritz, S.; Gyllenberg, M. A mechanistic derivation of the De Angelis-Beddington functional response. *J. Theor. Biol.* **2012**, *314*, 106–108. [CrossRef] [PubMed]
13. Geritz, S.A.H.; Gyllenberg, M. Group defence and the predator's functional response. *J. Math. Biol.* **2013**, *66*, 705–717. [CrossRef] [PubMed]
14. Geritz, S.A.H.; Gyllenberg, M. The De Angelis-Beddington functional response and the evolution of timidity of the prey. *J. Theor. Biol.* **2014**, *359*, 37–44. [CrossRef] [PubMed]
15. Metz, J.A.; Diekmann, O. *The Dynamics of Physiologically Structured Populations*; Springer: Berlin, Germany, 1986; Volume 68.
16. Rosenzweig, M.L.; MacArthur, R.H. Graphical representation and stability conditions of predator-prey interactions. *Am. Nat.* **1963**, *97*, 209–223. [CrossRef]
17. Gause, G.F. *The Struggle for Existence*; Williams and Wilkins: Baltimore, MD, USA, 1934.
18. Gause, G.F.; Smaragdova, N.P.; Witt, A.A. Further studies of interaction between predators and prey. *J. Anim. Ecol.* **1936**, *5*, 1–18. [CrossRef]
19. Dieckmann, U.; Doebeli, M. On the origin of species by sympatric speciation. *Nature* **1999**, *400*, 354–357. [CrossRef] [PubMed]
20. Bolnick, D.I. Intraspecific competition favours niche width expansion in Drosophila melanogaster. *Nature* **2001**, *410*, 463–466. [CrossRef] [PubMed]
21. Svanbäck, R.; Bolnick, D.I. Intraspecific competition drives increased resource use diversity within a natural population. *Proc. R. Soc. B Biol. Sci.* **2007**, *274*, 839–844. [CrossRef]
22. Araújo, M.S.; Guimaraes, P.R., Jr.; Svanbäck, R.; Pinheiro, A.; Guimarães, P.; Reis, S.F.D.; Bolnick, D.I. Network analysis reveals contrasting effects of intraspecific competition on individual vs. population diets. *Ecology* **2008**, *89*, 1981–1993. [CrossRef]

23. Bolnick, D.I. Can intraspecific competition drive disruptive selection? An experimental test in natural populations of sticklebacks. *Evolution* **2004**, *58*, 608–618. [CrossRef] [PubMed]
24. Kumar, S.; Kumar, R.; Cattani, C.; Samet, B. Chaotic behaviour of fractional predator-prey dynamical system. *Chaos Solitons Fract.* **2020**, *135*, 109811. [CrossRef]
25. Butusov, D.; Ostrovskii, V.; Tutueva, A.; Savelev, A. Comparing the algorithms of multiparametric bifurcation analysis. In Proceedings of the 2017 XX IEEE International Conference on Soft Computing and Measurements (SCM), St. Petersburg, Russia, 24–26 May 2017; pp. 194–198. [CrossRef]
26. Lehtinen, S.O.; Geritz, S.A. Cyclic prey evolution with cannibalistic predators. *J. Theor. Biol.* **2019**, *479*, 1–13. [CrossRef] [PubMed]
27. Lehtinen, S.O.; Geritz, S.A. Coevolution of cannibalistic predators and timid prey: evolutionary cycling and branching. *J. Theor. Biol.* **2019**, *483*, 110001. [CrossRef] [PubMed]

Alzheimer Identification through DNA Methylation and Artificial Intelligence Techniques

Gerardo Alfonso Perez [1,*] and Javier Caballero Villarraso [1,2]

[1] Department of Biochemistry and Molecular Biology, University of Cordoba, 14071 Cordoba, Spain; bc2cavij@uco.es
[2] Biochemical Laboratory, Reina Sofia University Hospital, 14004 Cordoba, Spain
* Correspondence: ga284@cantab.net

Abstract: A nonlinear approach to identifying combinations of CpGs DNA methylation data, as biomarkers for Alzheimer (AD) disease, is presented in this paper. It will be shown that the presented algorithm can substantially reduce the amount of CpGs used while generating forecasts that are more accurate than using all the CpGs available. It is assumed that the process, in principle, can be non-linear; hence, a non-linear approach might be more appropriate. The proposed algorithm selects which CpGs to use as input data in a classification problem that tries to distinguish between patients suffering from AD and healthy control individuals. This type of classification problem is suitable for techniques, such as support vector machines. The algorithm was used both at a single dataset level, as well as using multiple datasets. Developing robust algorithms for multi-datasets is challenging, due to the impact that small differences in laboratory procedures have in the obtained data. The approach that was followed in the paper can be expanded to multiple datasets, allowing for a gradual more granular understanding of the underlying process. A 92% successful classification rate was obtained, using the proposed method, which is a higher value than the result obtained using all the CpGs available. This is likely due to the reduction in the dimensionality of the data obtained by the algorithm that, in turn, helps to reduce the risk of reaching a local minima.

Keywords: algorithm; identification; Alzheimer

1. Introduction

Alzheimer (AD) is a relatively common neurological disorder associated with a decline in cognitive skills [1,2] and memory [3–5]. The causes of Alzheimer are not yet well understood, even as some processes of the development of amyloid plaque seems to be a major part of the disease [6]. The development of biomarkers [7] for the detection of AD is of clear importance. Over the last few decades, there has been a sharp increase in the amount of information publicly available, with researchers graciously making their data public. This, coupled with advances, such as the possibility to simultaneously estimate the methylation [8] levels of thousands of CpGs in the DNA, has created a large amount of information. CpG refers to having a guanine nucleotide after a cytosine nucleotide in a section of the DNA sequence. CpGs can be methylated, i.e., having an additional methyl group added. The level of methylation in the DNA is a frequently used marker for multiple illnesses [9–12], as well as a estimator of the biological age of the patient; hence, it has become an important biomarker [13]. The computational task is rather challenging. Current equipment can quickly analyze the level of methylation of in excess of 450,000 CpGs [14–16], with the latest generation of machines able to roughly double that amount [17]. As previously mentioned, methylation data has been linked to many diseases [18–20] and it is a logical research area for AD biomarkers. An additional challenge is that, at least in principle, there could be a highly non-linear process that is not necessarily accurately described by traditional regression analysis. The scope would then, hence, be to try to identify techniques that select a combination of the CpGs to be analyzed and then a non-linear algorithm that is able to predict whether the patient analyzed has the disease. However, on the other hand, it would not appear reasonable to totally discard the information presented in linear analysis. In the following sections, a mixed approach is presented. It will be shown that the approach is able to generate predictions (classifications between the control and patients suffering from Alzheimer).

1.1. Forecasting and Classification Models

Prediction and/or classification tasks are frequently found in many scientific and engineering fields with a large amount of potential artificial intelligence related techniques. The specific topics covered are rather diverse, including weather forecasts [21], plane flight time deviation [22], distributed networks [23], and many others [24–26]. One frequently used set of techniques are artificial neural networks. These techniques are extensively used in many fields. There are, however, several alternatives, which have received less attention in the existing literature (for instance, k-nearest neighbors and support vector machines). It should be noted that the k-nearest neighbor technique is frequently used in data pre-processing for instance in situations, in which the dataset has some missing values and the researcher needs to estimate those (typically as a previous step before using them as an input into a more complex model).

In our case the non-linear basic classification algorithm chosen was support vector machines (SVM) [27–29]. The basic idea of SVM is dividing the data into hyperplanes [30] and trying to decrease the measures of the classification error. This is achieved by following the usual supervised learning, in which a proportion of the data are used for training the SVM, while other portion (not used during the training phase) is used for testing purposes only, in order to avoid to avoid the issue of overfitting [5,31]. This technique has been applied in the context of Alzheimer for the classification of MRI images [32,33]. Some SVM models have been proposed in the context of CpGs methylation related to AD [34].

1.2. CpG DNA Methylation

A CpG is a dinucleotide pair (composed by cytosine a phosphate and guanine), while methylation refers to the addition of a methyl group to the DNA. Methylation levels are typically expressed as a percentage with 0 indicating completely unmethylated and 1 indicating 100% methylated. CpG DNA methylation levels are frequently used as epigenetic biomarkers [35,36]. Methylation levels change as an individual ages and this has been used to build biological clocks [37]. Individuals with some illnesses such as some cancers and Alzheimer present deviations in their levels of methylations.

1.3. Paper Structure

In the next section a related literature review is carried out given an overview of articles in prediction and classification. The literature review is followed by the materials and methods section, in which the main algorithm is explained. In this section, there is also a subsection describing the analyzed data. In Section 4 the results are presented. This section is divided into two subsection the first one describing the results for a single dataset and the second subsection describing the results when a multi dataset approach is followed. The last two sections are the discussion and the conclusions.

2. Literature Review

As previously mentioned, the CpG DNA methylation data were used in a variety of biomedical applications, such as the creation of biological clocks. For instance, Horvath [38] created an accurate CpG DNA methylation clock. Horvath managed to reduce the dimensionality of the data from hundred of thousands of CpGs analyzed per patient to a few hundred. This biological clock is able to predict the age of patients (in years) with rather high accuracy using as inputs the methylation data of a few hundred CpGs. A related article is [39], in which the authors used neural networks to predict the forensic age of individuals. The authors showed how using machine learning techniques could improve the accuracy of the age forecast, compared to traditional (linear) models.

Park et al. [40] is an interesting article focusing on DNA methylation and AD. The authors of this article found a link between DNA methylation and AD but similar to Horvath paper did not use machine learning techniques. Machine learning techniques have been applied with some success. For instance, ref. [41] used neural networks to analyze the relationship between gene-promoters methylation and biomarkers (one carbon metabolism in patients). Another interesting model was created by [42]. In this model the authors use a combination of DNA methylation and gene expression data to predict AD. The approached followed by the authors in this paper is different from the one that we pursued as they increased the amount of input data (including gene expression), while we focus on trying to reduce the dimensionality of the existing data i.e., select CpGs.

While most of the existing literature focuses on neural networks, there are also some interesting applications of other techniques such as for instance support vector machines (SVM). For instance, ref. [43] used SVM for the classification of histones. SVM have also been used for classification purposes in some illnesses such as colorectal cancer [44]. Even if SVM appears to be a natural choice for classification problems there seems to be less existing literature applying it to DNA methylation data in the context of AD identification.

3. Materials and Methods

One of the main objectives of this paper is to be able to accurately generate classification forecasts differentiating between individuals with Alzheimer's disease (AD) and control cases. The algorithm was built with the intention to be easily expandable from one to multiple data sets. A categorical variable y_i was created to classify individuals.

$$y_j = \begin{cases} 0 \text{ if } Control \\ 1 \text{ if } AD \end{cases} \quad (1)$$

In this way, a vector $Y = \{Y_1, Y_2, \ldots, Y_{nc}\}$ can be constructed classifying all the existing cases according to the disease estate (control or AD). In this notation nc denotes the total number, including both control and AD, of cases considered. Every case analyzed (j) has an associated vector X^j containing all the methylation levels of each CpG.

$$X^j = \begin{Bmatrix} X^1 \\ X^2 \\ \cdot \\ \cdot \\ \cdot \\ X^{mn} \end{Bmatrix} \quad (2)$$

This notation is used in order to clearly differentiate between the vector (X_j) containing all the methylation data for a single individual (all CpGs) from the vector (X_i) containing all the cases for a given CpG.

$$X_i = \{X_1, X_2, \ldots, X_{nc}\} \quad (3)$$

In a matrix notation the complete methylation data can be expressed as follows

$$X = \begin{pmatrix} X_1^1 & X_2^1 & \ldots & X_{nc}^1 \\ X_1^2 & X_2^2 & \ldots & X_{nc}^2 \\ \cdot & \cdot & & \cdot \\ \cdot & \cdot & & \cdot \\ \cdot & \cdot & & \cdot \\ X_1^{mn} & X_2^{mn} & \ldots & X_{nc}^{mn} \end{pmatrix} \quad (4)$$

For clarity purposes it is perhaps convenient shoving a hypothetical (oversimplified) example, in which 4 patients ($nc = 4$) are analyzed (2 control and 2 AD) and that only 5 CpGs were included per patient ($mn = 5$). In this hypothetical example:

$$Y = \{0, 0, 1, 1\} \tag{5}$$

As an example, the methylation data for patient 1 could be:

$$X^1 = \begin{Bmatrix} 0.9832 \\ 0.6145 \\ 0.1254 \\ 0.7845 \\ 0.6548 \end{Bmatrix} \tag{6}$$

Similarly, the methylation data for a single CpG for all patients can be expressed as:

$$X_i = \{0.9832, 0.3215, 0.6574, 0.6584\} \tag{7}$$

And the methylation data for all patients (matrix form) would be as follows:

$$X = \begin{pmatrix} 0.9832 & 0.3215 & 0.6574 & 0.6584 \\ 0.6145 & 0.6548 & 0.8475 & 0.7487 \\ 0.1254 & 0.6587 & 0.3254 & 0.6514 \\ 0.7845 & 0.3514 & 0.6254 & 0.6584 \\ 0.6548 & 0.6547 & 0.6587 & 0.6555 \end{pmatrix} \tag{8}$$

The proposed algorithm has two distinct steps. In the first step an initial filtering is carried out. This step reduced the dimensionality of the problem. The second step is the main algorithm. Both steps are described in the following subsections.

3.1. Initial Filtering

1. $\forall X_i$ estimate a linear regression with Y as the dependent variable. Save the p-value for each X_i.
2. Filter off the X_i with (p-value) < 0.005.

$$\{X_1, X_2, \ldots, X_{mn}\} \to \{X_1, X_2, \ldots, X_m\} \tag{9}$$

with $m < mn$.

3.2. Main Algorithm

1. Create a vector grid (D) with the each component representing the dimension (group of X_i) includes in the simulation. Two grids are included, a fine grid with relative small differences in the values of the elements (representing the dimensions that the researcher considers more likely) and a broad grid with large differences in values.

$$\text{Fine grid} = \{n_1, n_1 + \Delta n_s, n_1 + 2\Delta n_s, \ldots, n_1 + l\Delta n_s\} \tag{10}$$

$$\begin{aligned}\text{Broad grid} = \{&(n_1 + l\Delta n_s) + \Delta n_l, (n_1 + l\Delta n_s) + 2\Delta n_l, \ldots \\ &(n_1 + l\Delta n_s) + p\Delta n_l\}.\end{aligned} \tag{11}$$

The values inside the above grids represent the X_i selected. As an example, n_1 represents X_1. Δn_l and Δn_s are the constant step increases in the fine and broad grids, respectively. For instance, $n_1 + \Delta n_l$ and $n_1 + 2\Delta n_l$ are the second and third elements in the fine grid. The actual X_i elements related to this second and third values depend on the actual value of Δn_l. If $\Delta n_l = 1$ then the second and third elements related to X_2 and X_3, respectively, while if $\Delta n_l = 2$, then they relate to X_3 and X_5, respectively. Where $\Delta n_l > \Delta n_s$, each of these values, i.e., $n_1 + \Delta n_s$ is the number of x_i chosen. $l \in \mathbb{Z}^+$ is a constant that specifies (together with n_l) the total size of the fine grid, while $p \in \mathbb{Z}^+$ is the analogous term for the broad grid. For simplicity purposes the case of a fine grid, starting a X_1, followed by a broad grid has been shown but this is not a required constraint. The intent is giving discretion to the researcher to apply the fine grid to the area that is considered more important. This is an attempt to bring the expertise of the researcher into the algorithm. In Equation (12) it can be seen the combination of these two grids (D).

$$D = \{n_1, n_1 + \Delta n_s, n_1 + 2\Delta n_s, \ldots, n_1 + l\Delta n_s, (n_1 + l\Delta n_s) + \Delta n_l, \\ (n_1 + l\Delta n_s) + 2\Delta n_l, \ldots, (n_1 + l\Delta n_s) + p\Delta n_l\}. \tag{12}$$

For clarity purposes, let simplify the notation:

$$D = \{S_j\} = \{S_1, S_2, \ldots, S_m\} \tag{13}$$

where Equations (12) and (13) are identical. "S" is a more compact notation with for instance S_1 and S_2 representing n_1 and $n_1 + \Delta n_s$, respectively.

2. Create a mapping between each $x_i = \{X_1, \ldots, X_m\} = \{X_i\}$, where each X_i is a vector, and 10 decile regions. The group of X_i with the highest 10% of the p-value are included in the first decile and assigned a probability of 100%. The group of X_i with the second highest 10% of the p-value are included in the second decile and assigned a probability of 90%. This process is repeated for all deciles creating a mapping.

$$\{X_1, \ldots, X_m\} \rightarrow B\{1.0, 0.9, 0.8, \ldots, 0.1\} \tag{14}$$

Where B is a vector of probabilities. In this way, the X_i with the largest p-values are more likely to be included.

3. For each S_j generate $\forall X_i$, i=1,...,m, a random number R_i with ($0 \leq R_i \leq 1$). If $R_i > B\{X_i\}$ then X_i is not included in the preliminary S_j group of X_is. Otherwise it is included. In this way a filtering is carried out.

$$\{X_1, \ldots, X_m\} \rightarrow \{X_1, \ldots, X_{m*}\} \forall S_j \tag{15}$$

4. Randomly S_j elements of m^* are chosen.
5. Estimate the Hit Ratio (HR)

$$HR = \frac{CE}{TE} \tag{16}$$

where TE is the total number of classification estimations and CE is the number of correct classification estimates.

6. Repeat steps (3) to (6) k times for each S_j. In this way there is a mapping:

$$\{S_1, \ldots, S_m\} \rightarrow \{HR(S_1), \ldots, HR(S_m)\} \tag{17}$$

Remark 1. *An alternative approach would be choosing the starting distribution S_j as the one after which the mean value of the HR does not statistically increase at a 5% confidence level.*

7. Define new search interval between the two highest success rates:

$$max\{HR(S_1), \ldots, HR(m)\} \rightarrow S^1_{max} \tag{18}$$

$$max\{HR(S_1), \ldots, HR(m)\} < S^1_{max} \rightarrow S^1_{max-1} \tag{19}$$

Iteration 1 (Iter=1) ends, identifying interval:

$$\{S^1_{max}, S^1_{max-1}\} \tag{20}$$

Remark 2. *It is assumed, for simplicity, without loss of generality that $S^1_{max} < S^1_{max-1}$. If that it is not the case then the interval needs to be switched ($\{S^1_{max-1}, S^1_{max}\}$).*

8. Divide the interval identified in the previous step into $k-1$ steps.

$$\{S_1, \ldots, S_k\} \qquad (21)$$

where $S_1 = S^1_{max}$ and $S_k = S^1_{max-1}$

9. Create a new mapping estimating the new hit rates (following the same approach as in previous steps)

$$\{S_1, \ldots, S_k\} = \{HR(S_1), \ldots, HR(S_k)\} \qquad (22)$$

10. Repeat $Iter_t$ times until the maximum number of iterations ($Iter_{max}$) is reached.

$$Iter_t \geq Iter_{max} \qquad (23)$$

or until the desire hit rate ($HR_{desired}$) is reached

$$HR(S) \leq HR_{desired} \qquad (24)$$

or until no further HR improvement is achieved. Select S^t_{max}.

A few points need to be highlighted. It is important to reduce the number of combinations to a manageable size. For instance, assuming that there are "m" X_i (after the initial filtering of p-Values) there would be $\binom{m}{r}$ combinations of size r. The well known equation (25) can be used.

$$\sum_{r=0}^{m} \binom{m}{r} = 2^m \; \forall m \in \mathbf{N}^+ \qquad (25)$$

Assuming that at least one of the X_i is selected:

$$\sum_{r=0}^{m} \binom{m}{r} = \sum_{r=1}^{m} \binom{m}{r} + \binom{m}{0} = 2^m \qquad (26)$$

$$\sum_{r=1}^{m} \binom{m}{r} = 2^m - 1 \qquad (27)$$

For large m values the -1 term is negligible.

In the initial step the problem of having to calculate the estimations for 2^m combinations is simplified into calculating a $q2^q$ combinations with $q < m$. If for example, $q = m/10$, then the problem is reduced form 2^{10q} to $10 2^q$ combinations. It can be proven that:

$$2^{10q} > 10 \cdot 2^q \; \forall q \geq 2 \qquad (28)$$

Proof. Using induction. Base case (q=2). $2^{10(k)} = 2^{20} = 1,048,576$; $10 \cdot 2^q = 10 \cdot 2^2 = 40$. $1,048,576 > 40$. Therefore, the base case is confirmed. Assume:

$$2^{10k} > 10 \cdot 2^k \text{ for some } k \geq 2 \qquad (29)$$

induction hypothesis

$$2^{10(k+1)} > 10 \cdot 2^{k+1} \qquad (30)$$

$$2^{10(k+1)} = 2^{10k} 2^{10} > 10 \cdot 2^k 2^{10} = 10 \cdot 2^k 2^2 9 = 10 \cdot 2^{k+1} 2^9 > 10 \cdot 2^{k+1} \qquad (31)$$

which completes the proof by induction. □

3.3. Data

The methylation data set (Table 1) were obtained from the GEO database and the corresponding accession codes are shown in the table. The methylation data in these two experiments was obtained following similar approaches and both experiments used an Illumina machine. The raw data were structured in a matrix form. For clarity purposes a sample for an specific individual is shown in Table 2. In this table it can be seen the methylation level for all 481,868 CpGs analyzed for a single patient. In the second column it can be seen the identification number for each specific CpG, while in the third column the level of methylation for each specific CpG is shown. Please notice that this is a percentage value ranging from 0 (no methylation) to 1 (fully methylated). Additionally, each patient in the database will be classified according to a binary variable showing if the patient has Alzheimer of if he/she is a healthy control individual. The binary classification variable can be seen in the last row of the table (it is either a 0 or a 1).

Table 1. Methylation data sets included in the analysis.

GEO Code	Cases	Tissue	Illness
GSE66351	190	Glian and neuron	AD and control
GSE80970	286	Pre-frontal cortex and gyrus	AD and control

Table 2. Single patient methylation data.

Number	CpG (Indetifier)	Methylation Level
1	cg13869341	0.89345
2	cg14008030	0.71088
...	...	...
481,868	cg05999368	0.51372
AD/Control		0

Hence, the problem becomes a classification problem, in which the algorithm has to identify how many and which CpGs to use in order to appropriately classify the individuals in the two categories (AD and healthy). A oversimplified sample (not accurate for classification purposes but rather clear for explanation purposes) is shown in Table 3. In this (unrealistic) case only two CpGs were selected for each patient.

Table 3. Single patient methylation data.

Number	CpG (Indetifier)	Methylation Level
2	cg14008030	0.71088
481,868	cg05999368	0.51372
AD/Control		0

It is perhaps easier to conceptualize if the number and the CpG identifier are omitted and several patients are shown (Table 4). This table shows the results (for illustration purposes only) of an unrealistic case, in which the algorithm selects only two CpGs for each patient. Three patient in total are shown, two are control patients and one has AD. This clearly illustrates the objective of the algorithm, which is Selectric the CpGs (rows in this notation) to classify each patient (columns in this notation) according to a binary variable (last row in this notation).

Table 4. Multiple patient methylation data.

Patient 1	Patient 2	Patient 3
0.71088	0.63174	0.72582
0.51372	0.62145	0.43212
0	1	0

In this notation, the Table 4 is the solution generated by the algorithm when presented with the original data of the form shown in Table 5. Table 5 shows all the potential input variables X_i^j (to be selected) where, as previously mentioned, "i" identifies all the potential CpGs per patient and the index "j" identifies the patient. The variable Y_i is the binary variable associated with each patient differentiating between healthy an AD individuals. When expressed in this notation, it is easy to see that the problem boils down to a classification problem, suitable for techniques such as support vector machines.

Table 5. Multiple patient methylation data (general data structure).

Patient 1	Patient 2	Patient 3
X_1^1	X_1^2	X_1^3
X_2^1	X_2^2	X_2^3
...	...	...
$X_{481,868}^1$	$X_{481,868}^2$	$X_{481,868}^3$
Y_1	Y_2	Y_3

4. Results

4.1. Single Data Set

Initially a first estimation using all the available CpGs and a support vector machine classifier was used. The age of the patient (Table 6) was one of the main factors affecting the accuracy of the patient classification using the data set GSE 66351. Controlling for age allowed for better HR rates. Controlling for other variables, such as gender, cell type, or brain region did not appear to improve the classification accuracy . Three different kernels were used (linear, Gaussian, and polynomial), with the best results obtained when using the linear kernel.

Table 6. Hit Rate (HR) of SVM with 3 different kernels for Alzheimer classification (versus control patients), using all the CpGs available (481,778) and controlling for different factors, such as age, gender, cell type, or brain region (GSE 66351 test data).

Controls	HR (Linear)	HR (Gaussian)	HR (Polynomial)	CpGs
None	0.8211	0.7921	0.8167	All
Age	0.8947	0.8142	0.8391	All
Gender	0.8211	0.7921	0.8167	All
Cell type	0.8211	0.7921	0.8167	All
Brain Region	0.8211	0.7921	0.8167	All

In the initial filtering stage the linear regression between each CpGs (X_i) and the vector classification (identifying patients suffering from Alzheimer and control patients was carried out and the p-values stored. CpGs with p-values higher than 0.05 were excluded. The remaining 41,784 CpGs were included in the analysis. It can be seen in Table 7 that as in the previous case controlling for age did improve the HR. The linear kernel was used.

Table 7. HR of SVM for Alzheimer classification (versus control patients), using all CpGs with p-values < 0.05 (41,784) and controlling for different factors, such as age, gender, cell type, or brain region (GSE 66351 test data).

Controls	Hit Rate	CpGs
None	0.7263	41,784
Age	0.8424	41,784
Gender	0.7263	41,784
Cell type	0.7263	41,784
Brain Region	0.7263	41,784

In Figure 1 it is shown that it is possible to achieve high HR using a subset of the CpGs. This HR is higher than the one obtained using all CpGs. As in all the previous cases, the HR rate showed is the out-of-sample HR, i.e., the HR obtained using the testing data that were not used during the training phase. The SVM was trained with approximately 50% of the data contained in the GSE 66351 data set. The testing and training datasets were divided in a manner that roughly maintained the same proportion of control and AD individuals in both datasets. 10-fold cross validation was carried out to try to ensure model robustness. The SVM used linear kernel. The analysis in this figure was carried out controlling for age, gender, cell type and brain region. As in previous cases, the only factor that appears to have an impact on the calculation, besides the level of methylation of the CpGs, was the age. In total, 190 cases of this database was used for either training or testing purposes. The maximum HR obtained was 0.9684, obtained while using 1000 CpGs.

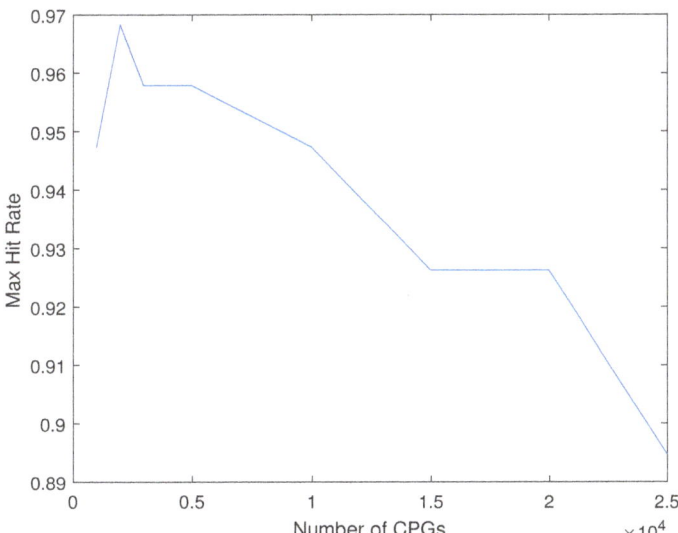

Figure 1. Max Hit Rate (HR) versus number of CpGs included in the analysis.

Figure 2 shows the alternative approach mentioned in the methodology, rather than the maximum HR rate obtained the figure shows the average HR obtained at each level(number of CpGS) and its related confidence interval (5%). It is clear from both Figures 1 and 2 that regardless of the approach followed it appears that after a certain amount of CpGs adding additional CpGs to the analysis does not further increased the HR.

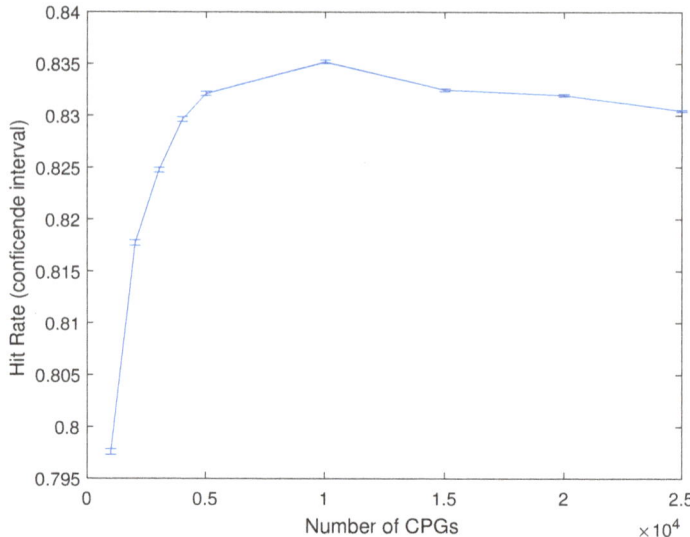

Figure 2. Average Hit Rate (HR) and confidence interval (5%) versus number of CpGs included in the analysis.

4.2. Multiple Data Sets

One of the practical issues when carrying out this type of analysis is the lack of consistency between databases, even when there are following similar empirical approaches. As an example, in the case of the GSE66351 dataset a total of 41,784 CpGs were found to be statistically significant (after data pre-processing). Of these 41,784 CpGs only 18.98% (7929) were found to be statistically significant (same p-value) in the GSE80970 dataset. This is likely due to subtle different in experimental procedures. In order to overcome this issue only the 7929 CpGs statistically significant CpGs were used when analyzing these two combined datasets. Besides this different pre-filtering step the rest of the algorithm used was as described in the previous section. Both data sets were combined and divided into a training and a test data set.

One of the main differences in the results, besides the actual HR, is that including the age of the patient in the algorithm (using these reduced starting CpG pools) did not appear to substantially increase the forecasting accuracy of the model. The best results when using this approach were obtained when using 4300 CpGs with a combined HR (out of sample) of 0.9202 (Table 8). The list of the 4300 CpGs can be found in the supplementary material.

Table 8. HR of SVM for AD vs. control patients using 4300 CpGs.

Controls	Hit Rate	CpGs
GSE66351	0.8710	4300
GSE80970	0.9517	4300
All	0.9202	4300

Following the standard practice [45] the sensitivity, specificity, positive predictive value (PPV) and negative predictive ratio (NPV) were calculated for all the testing data combined as well as for the testing data in the GSE66351 and GSE80970 separately, Table 9, using the obtained model (4300 CpGs) All the cases included in the analysis are out-of-sample cases, i.e., not previously used during the training of the support vector machine. It is important to obtain models that are able to generalize well across different data sets.

Table 9. Classification ratios (out-of-sample), including positive predictive value (PPV) and negative predictive ratio (NPV).

Ratio	All	GSE66351	GSE80970
Sensitivity	0.9007	0.8333	0.9506
Specificity	0.9485	0.9394	0.9531
PPV	0.9621	0.9615	0.9625
NPV	0.8679	0.7561	0.9385

5. Discussion

In this paper, an algorithm for the selection of DNA methylation CpG data is presented. A substantial reduction on the number of CpGs analyzed is achieved, while the classification precision is higher than when using all CpGs available. The algorithm is designed to be scalable. In this way, as more data set of Alzheimer DNA methylation become available, the analysis can be gradually expanding. There appear to be substantial differences in the data contained in the data sets analyzed. This is likely due to relatively small experimental procedures. There results obtained (two data sets) are reasonably precise with a sensitivity of 0.9007 and a specificity of 0.9485, while the PPV and the NPV were 0.9621 and 0.8679, respectively. It was also appreciated that when using large amounts of CpGs controlling for age was a crucial steps. However, as the number of CpGs selected by the algorithm decreased, the importance of controlling for age also decreased. Given the large amount of possible combinations of CpGs it is of clear importance to develop algorithm for their selection. As an example, it is clearly not feasible to calculate all the possible combinations of a data set composed by 450,000 CpGs.

The results highlight the necessity to reduce the dimensionality of the data. This is not only in order to facilitate the computations but from a purely statistical point of view, as well. Ideally the number of factors considered should be of the same order of magnitude than the number of samples. In this situation there is a large amount of factors (+450,000) per individual but a relatively small number of individuals. Besides some very specific trails, such as the ongoing SARS-CoV-2 (COVID-19) trials of some vaccines, it is very unlikely to have a cohort of patients and control individuals approaching 450,000. The accuracy of the forecasts increases when the dimensionality of the data are reduced. This is likely due to a reduction of the risk of the algorithm reaching a local minima.

Several methodological decisions were made in order to try to improve the generalization power of the model, i.e., the ability to generate accurate forecast when faced with new data. One of this decisions was to have a large (50%) testing dataset and to have a process that can accommodate for multiple datasets as they become available.

6. Conclusions

Having techniques that can determine if an individual has Alzheimer disease is likely going to become increasingly important. This area of research has, arguably, not received enough attention in the past. This is probably due to the fact that there was no treatment available. This has recently changed, with the FDA approving [46–49] the first drug for the treatment of Alzheimer disease (there were drugs before targeting some of the effects of the illness but not the actual illness itself).

The results, for instance, in Table 9, suggest that the approached followed can generate an accurate forecast (out-of-sample), when using a multi dataset approach, which is a significant development, with, for instance, the sensitivity and the specificity reaching, respectively, 0.9007 and 0.9485 values, when using 4300 CpGs. The obtained positive predictive value (PPV) and the negative predictive value (NPV) were also relatively high, coming in at 0.9621 and 0.8679, respectively. The results also indicate (Figures 1 and 2) that increasing the number of CpGs does not improve the forecast. This is very likely related to the issue of local minima.

It is also important to remark that, as more data becomes available, the algorithm could be used to classify between healthy and AD patients following a less invasive approach. Most of the currently available methylation data are related to brain tissue that requires an invasive procedure to be obtained. However, methylation datasets in numerous other illnesses already exist, using blood. As blood-based datasets become available, the algorithm presented in this paper can be easily applied to those, potentially becoming an additional practical tool for diagnosis of the illness. There are also several interesting lines of future work. For instance, the addition of new datasets as they become gradually available.

Supplementary Materials: The following are available online at https://www.mdpi.com/2227-7390/9/19/2482/s1.

Author Contributions: Conceptualization, G.A.P.; methodology, G.A.P. and J.C.V.; software, G.A.P.; validation, G.A.P. and J.C.V.; formal analysis, G.A.P. and J.C.V.; investigation, G.A.P. and J.C.V.; resources, G.A.P. and J.C.V.; data curation, G.A.P. and J.C.V.; writing—original draft preparation, G.A.P.; writing—review and editing, G.A.P. and J.C.V.; visualization, G.A.P. and J.C.V.; supervision, G.A.P. and J.C.V.; project administration, G.A.P. and J.C.V.; funding acquisition, G.A.P. and J.C.V. All authors have read and agreed to the published version of the manuscript.

Funding: This research received no external funding.

Data Availability Statement: All the data used in this paper is publicly available at the GEO Database (https://www.ncbi.nlm.nih.gov/geo/, accessed on 1 July 2021).

Conflicts of Interest: The authors declare no conflict of interest.

References

1. Olivari, B.S.; Baumgart, M.; Taylor, C.A.; McGuire, L.C. Population measures of subjective cognitive decline: A means of advancing public health policy to address cognitive health. *Alzheimer's Dement. Transl. Res. Clin. Interv.* **2021**, *7*, e12142.
2. Donohue, M.C.; Sperling, R.A.; Salmon, D.P.; Rentz, D.M.; Raman, R.; Thomas, R.G.; Weiner, M.; Aisen, P.S. The preclinical Alzheimer cognitive composite: Measuring amyloid-related decline. *JAMA Neurol.* **2014**, *71*, 961–970. [CrossRef]
3. Morris, R.G.; Kopelman, M.D. The memory deficits in Alzheimer-type dementia: A review. *Q. J. Exp. Psychol.* **1986**, *38*, 575–602. [CrossRef]
4. Greene, J.D.; Hodges, J.R.; Baddeley, A.D. Autobiographical memory and executive function in early dementia of Alzheimer type. *Neuropsychologia* **1995**, *33*, 1647–1670. [CrossRef]
5. Sahakian, B.J.; Morris, R.G.; Evenden, J.L.; Heald, A.; Levy, R.; Philpot, M.; Robbins, T.W. A comparative study of visuospatial memory and learning in Alzheimer-type dementia and Parkinson's disease. *Brain* **1988**, *111*, 695–718. [CrossRef] [PubMed]
6. Serrano-Pozo, A.; Frosch, M.P.; Masliah, E.; Hyman, B.T. Neuropathological alterations in Alzheimer disease. *Cold Spring Harb. Perspect. Med.* **2011**, *1*, a006189. [CrossRef] [PubMed]
7. Blennow, K.; Hampel, H.; Weiner, M.; Zetterberg, H. Cerebrospinal fluid and plasma biomarkers in Alzheimer disease. *Nat. Rev. Neurol.* **2010**, *6*, 131–144. [CrossRef]
8. Hsieh, C.L. Dependence of transcriptional repression on CpG methylation density. *Mol. Cell. Biol.* **1994**, *14*, 5487–5494. [CrossRef]
9. Cooper, D.N.; Krawczak, M. Cytosine methylation and the fate of CpG dinucleotides in vertebrate genomes. *Hum. Genet.* **1989**, *83*, 181–188. [CrossRef]
10. Vertino, P.M.; Yen, R.; Gao, J.; Baylin, S.B. De novo methylation of CpG island sequences in human fibroblasts overexpressing DNA (cytosine-5-)-methyltransferase. *Mol. Cell. Biol.* **1996**, *16*, 4555–4565. [CrossRef]
11. Gudjonsson, J.E.; Krueger, G. A role for epigenetics in psoriasis: Methylated cytosine–guanine sites differentiate lesional from nonlesional skin and from normal skin. *J. Investig. Dermatol.* **2012**, *132*, 506–508. [CrossRef] [PubMed]
12. Cornélie, S.; Wiel, E.; Lund, N.; Lebuffe, G.; Vendeville, C.; Riveau, G.; Vallet, B.; Ban, E. Cytosine-phosphate-guanine (CpG) motifs are sensitizing agents for lipopolysaccharide in toxic shock model. *Intensive Care Med.* **2002**, *28*, 1340–1347. [CrossRef] [PubMed]
13. Mikeska, T.; Craig, J.M. DNA methylation biomarkers: Cancer and beyond. *Genes* **2014**, *5*, 821–864. [CrossRef]
14. Pidsley, R.; Wong, C.C.; Volta, M.; Lunnon, K.; Mill, J.; Schalkwyk, L.C. A data-driven approach to preprocessing Illumina 450K methylation array data. *BMC Genom.* **2013**, *14*, 293. [CrossRef]
15. Marabita, F.; Almgren, M.; Lindholm, M.E.; Ruhrmann, S.; Fagerström-Billai, F.; Jagodic, M.; Sundberg, C.J.; Ekström, T.J.; Teschendorff, A.E.; Tegnér, J.; et al. An evaluation of analysis pipelines for DNA methylation profiling using the Illumina HumanMethylation450 BeadChip platform. *Epigenetics* **2013**, *8*, 333–346. [CrossRef] [PubMed]

16. Kuan, P.F.; Wang, S.; Zhou, X.; Chu, H. A statistical framework for Illumina DNA methylation arrays. *Bioinformatics* **2010**, *26*, 2849–2855. [CrossRef]
17. You, L.; Han, Q.; Zhu, L.; Zhu, Y.; Bao, C.; Yang, C.; Lei, W.; Qian, W. Decitabine-mediated epigenetic reprograming enhances anti-leukemia efficacy of CD123-targeted chimeric antigen receptor T-cells. *Front. Immunol.* **2020**, *11*, 1787. [CrossRef]
18. Rhee, I.; Jair, K.W.; Yen, R.W.C.; Lengauer, C.; Herman, J.G.; Kinzler, K.W.; Vogelstein, B.; Baylin, S.B.; Schuebel, K.E. CpG methylation is maintained in human cancer cells lacking DNMT1. *Nature* **2000**, *404*, 1003–1007. [CrossRef]
19. Feng, W.; Shen, L.; Wen, S.; Rosen, D.G.; Jelinek, J.; Hu, X.; Huan, S.; Huang, M.; Liu, J.; Sahin, A.A.; et al. Correlation between CpG methylation profiles and hormone receptor status in breast cancers. *Breast Cancer Res.* **2007**, *9*, 1–13. [CrossRef]
20. Lin, R.K.; Hsu, H.S.; Chang, J.W.; Chen, C.Y.; Chen, J.T.; Wang, Y.C. Alteration of DNA methyltransferases contributes to 5 CpG methylation and poor prognosis in lung cancer. *Lung Cancer* **2007**, *55*, 205–213. [CrossRef]
21. Haupt, S.E.; Cowie, J.; Linden, S.; McCandless, T.; Kosovic, B.; Alessandrini, S. Machine learning for applied weather prediction. In Proceedings of the 2018 IEEE 14th International Conference on e-Science (e-Science), Amsterdam, The Netherlands, 29 October–1 November 2018; pp. 276–277.
22. Stefanovič, P.; Štrimaitis, R.; Kurasova, O. Prediction of flight time deviation for lithuanian airports using supervised machine learning model. *Comput. Intell. Neurosci.* **2020**. [CrossRef]
23. Rafiee, P.; Mirjalily, G. Distributed Network Coding-Aware Routing Protocol Incorporating Fuzzy-Logic-Based Forwarders in Wireless Ad hoc Networks. *J. Netw. Syst. Manag.* **2020**, *28*, 1279–1315. [CrossRef]
24. Roshani, M.; Phan, G.; Roshani, G.H.; Hanus, R.; Nazemi, B.; Corniani, E.; Nazemi, E. Combination of X-ray tube and GMDH neural network as a nondestructive and potential technique for measuring characteristics of gas-oil–water three phase flows. *Measurement* **2021**, *168*, 108427. [CrossRef]
25. Pourbemany, J.; Essa, A.; Zhu, Y. Real Time Video based Heart and Respiration Rate Monitoring. *arXiv* **2021**, arXiv:2106.02669.
26. Alfonso, G.; Carnerero, A.D.; Ramirez, D.R.; Alamo, T. Stock forecasting using local data. *IEEE Access* **2020**, *9*, 9334–9344. [CrossRef]
27. Joachims, T. *SVM-Light: Support Vector Machine*, version 6.02; University of Dortmund: Dortmund, Germany, 1999.
28. Meyer, D.; Leisch, F.; Hornik, K. The support vector machine under test. *Neurocomputing* **2003**, *55*, 169–186. [CrossRef]
29. Wang, L. *Support Vector Machines: Theory and Applications*; Springer Science & Business Media: Berlin/Heidelberg, Germany, 2005; Volume 177.
30. Noble, W.S. What is a support vector machine? *Nat. Biotechnol.* **2006**, *24*, 1565–1567. [CrossRef]
31. Li, X.; Wang, L.; Sung, E. A study of AdaBoost with SVM based weak learners. In Proceedings of the 2005 IEEE International Joint Conference on Neural Networks, Montreal, QC, Canada, 31 July–4 August 2005; Volume 1, pp. 196–201.
32. Magnin, B.; Mesrob, L.; Kinkingnéhun, S.; Pélégrini-Issac, M.; Colliot, O.; Sarazin, M.; Dubois, B.; Lehéricy, S.; Benali, H. Support vector machine-based classification of Alzheimer's disease from whole-brain anatomical MRI. *Neuroradiology* **2009**, *51*, 73–83. [CrossRef] [PubMed]
33. Wang, S.; Lu, S.; Dong, Z.; Yang, J.; Yang, M.; Zhang, Y. Dual-tree complex wavelet transform and twin support vector machine for pathological brain detection. *Appl. Sci.* **2016**, *6*, 169. [CrossRef]
34. Fetahu, I.S.; Ma, D.; Rabidou, K.; Argueta, C.; Smith, M.; Liu, H.; Wu, F.; Shi, Y.G. Epigenetic signatures of methylated DNA cytosine in Alzheimer's disease. *Sci. Adv.* **2019**, *5*, eaaw2880. [CrossRef] [PubMed]
35. Tost, J. DNA methylation: An introduction to the biology and the disease-associated changes of a promising biomarker. *Mol. Biotechnol.* **2010**, *44*, 71–81. [CrossRef] [PubMed]
36. Rauch, T.A.; Wang, Z.; Wu, X.; Kernstine, K.H.; Riggs, A.D.; Pfeifer, G.P. DNA methylation biomarkers for lung cancer. *Tumor Biol.* **2012**, *33*, 287–296. [CrossRef] [PubMed]
37. Horvath, S.; Raj, K. DNA methylation-based biomarkers and the epigenetic clock theory of ageing. *Nat. Rev. Genet.* **2018**, *19*, 371–384. [CrossRef] [PubMed]
38. Horvath, S. DNA methylation age of human tissues and cell types. *Genome Biol.* **2013**, *14*, 1–20. [CrossRef]
39. Vidaki, A.; Ballard, D.; Aliferi, A.; Miller, T.H.; Barron, L.P.; Court, D.S. DNA methylation-based forensic age prediction using artificial neural networks and next generation sequencing. *Forensic Sci. Int. Genet.* **2017**, *28*, 225–236. [CrossRef] [PubMed]
40. Mastroeni, D.; Grover, A.; Delvaux, E.; Whiteside, C.; Coleman, P.D.; Rogers, J. Epigenetic changes in Alzheimer's disease: Decrements in DNA methylation. *Neurobiol. Aging* **2010**, *31*, 2025–2037. [CrossRef] [PubMed]
41. Grossi, E.; Stoccoro, A.; Tannorella, P.; Migliore, L.; Coppedè, F. Artificial neural networks link one-carbon metabolism to gene-promoter methylation in Alzheimer's disease. *J. Alzheimer's Dis.* **2016**, *53*, 1517–1522. [CrossRef]
42. Park, C.; Ha, J.; Park, S. Prediction of Alzheimer's disease based on deep neural network by integrating gene expression and DNA methylation dataset. *Expert Syst. Appl.* **2020**, *140*, 112873. [CrossRef]
43. Bhasin, M.; Reinherz, E.L.; Reche, P.A. Recognition and classification of histones using support vector machine. *J. Comput. Biol.* **2006**, *13*, 102–112. [CrossRef]
44. Zhao, D.; Liu, H.; Zheng, Y.; He, Y.; Lu, D.; Lyu, C. A reliable method for colorectal cancer prediction based on feature selection and support vector machine. *Med. Biol. Eng. Comput.* **2019**, *57*, 901–912. [CrossRef]
45. Lalkhen, A.G.; McCluskey, A. Clinical tests: Sensitivity and specificity. *Contin. Educ. Anaesth. Crit. Care Pain* **2008**, *8*, 221–223. [CrossRef]

46. Tanzi, R.E. FDA Approval of Aduhelm Paves a New Path for Alzheimer's Disease. *ACS Chem. Neurosci.* **2021**, *12*, 2714–2715. [CrossRef] [PubMed]
47. Karlawish, J.; Grill, J.D. The approval of Aduhelm risks eroding public trust in Alzheimer research and the FDA. *Nat. Rev. Neurol.* **2021**, *17*, 523–524. [CrossRef]
48. Ayton, S. Brain volume loss due to donanemab. *Eur. J. Neurol.* **2021**, *28*, e67–e68. [CrossRef]
49. Vellas, B.J. The Geriatrician, the Primary Care Physician, Aducanumap and the FDA Decision: From Frustration to New Hope. *J. Nutr. Health Aging* **2021**, *25*, 821–823. [CrossRef] [PubMed]

Article

Visual Sequential Search Test Analysis: An Algorithmic Approach

Giuseppe Alessio D'Inverno [1,*], Sara Brunetti [1], Maria Lucia Sampoli [1], Dafin Fior Muresanu [2,3], Alessandra Rufa [4] and Monica Bianchini [1]

1 Department of Information Engineering and Mathematics, University of Siena, 53100 Siena, Italy; sara.brunetti@unisi.it (S.B.); marialucia.sampoli@unisi.it (M.L.S.); monica.bianchini@unisi.it (M.B.)
2 Department of Neurosciences, "Iuliu Haţieganu" University of Medicine and Pharmacy, 400023 Cluj-Napoca, Romania; dafinm@ssnn.ro
3 RoNeuro Institute for Neurological Research and Diagnostic, 400364 Cluj-Napoca, Romania
4 Department of Medicine, Surgery and Neuroscience, University of Siena, 53100 Siena, Italy; rufa@unisi.it
* Correspondence: dinverno@diism.unisi.it

Abstract: In this work we present an algorithmic approach to the analysis of the Visual Sequential Search Test (VSST) based on the episode matching method. The data set included two groups of patients, one with Parkinson's disease, and another with chronic pain syndrome, along with a control group. The VSST is an eye-tracking modified version of the Trail Making Test (TMT) which evaluates high order cognitive functions. The episode matching method is traditionally used in bioinformatics applications. Here it is used in a different context which helps us to assign a score to a set of patients, under a specific VSST task to perform. Experimental results provide statistical evidence of the different behaviour among different classes of patients, according to different pathologies.

Keywords: visual sequential search test; episode matching; trail making test; sequence alignment; alignment score

Citation: D'Inverno, G.A.; Brunetti, S.; Sampoli, M.L.; Muresanu, D.F.; Rufa, A.; Bianchini, M. Visual Sequential Search Test Analysis: An Algorithmic Approach. *Mathematics* **2021**, *9*, 2952. https://doi.org/10.3390/math9222952

Academic Editor: Radu Tudor Ionescu

Received: 17 October 2021
Accepted: 17 November 2021
Published: 18 November 2021

Publisher's Note: MDPI stays neutral with regard to jurisdictional claims in published maps and institutional affiliations.

Copyright: © 2021 by the authors. Licensee MDPI, Basel, Switzerland. This article is an open access article distributed under the terms and conditions of the Creative Commons Attribution (CC BY) license (https://creativecommons.org/licenses/by/4.0/).

1. Introduction

The Trail Making Test (TMT) is a popular neuropsychological test, commonly used in clinical settings as a diagnostic tool for the evaluation of some frontal functions. It provides qualitative information on high order mental activities, including speed of processing, mental flexibility, visual spatial orientation, working memory and executive functions. Originally, it was part of the Army Individual Test Battery (1944) and subsequently was incorporated into the Halstead–Reitan Battery [1]. In general terms, the test consists of a visual search of objects in an image, where the objects are arranged in sequences of loci called regions of interest (ROIs). While classical TMT requires an individual to draw lines sequentially connecting an assigned sequence of letters and/or numbers (the ROIs) with a pencil or mouse, the same task can be performed by using the eye-tracking technology and asking the subject to fixate the sequence of ROIs in the prescribed order [2]. Poor performance is known to be associated with many types of brain impairment, in particular frontal lobe lesions. For instance, eye-tracking studies have proved their efficacy in the diagnosis of many common neurological pathologies, such as Parkinson's disease, brain trauma and neglect phenomena [3–6].

The Visual Sequential Search Test (VSST) adopts the same principle as the TMT but is based on a precise geometry of the ROIs, being designed for the study of top-down visual search. Visual search can be quantified in terms of the analysis of the scan-path, which is a sequence of saccades and fixations. Thus, the identification of precise scores of the VSST may provide a measure of the subject's visual spatial ability and high order mental activity. Specifically, we know that visual input from the external world is actively sampled for promoting appropriate actions during everyday life; this mechanism is dynamic and involves a continuous re-sampling of spatial elements of a visual scene. The VSST is a repeated search task, in which patients are asked to connect by gaze a logical sequence of numbers and letters.

The main objective of the research is to identify a reliable method for the analysis of the VSST and to investigate common/different characteristics inside and outside three different subjects' classes. The first group includes patients with extrapyramidal disease who have well known difficulties in visual spatial exploration and executive functions. Thus, we predict low VSS performance in this group of patients. The second group is composed of patients suffering from chronic pain syndrome. This syndrome is not classically associated with cognitive deficits, but rather with mood changes. However, a possible deficit of attention has been suggested in patients suffering from chronic pain. If the prediction is true, the VSS performance in this group should be normal or less altered than extrapyramidal patients. The third, is a control group. The identification of a robust method of analysis and the detection of a reliable indicator for the VSST performance would allow to give a measure of executive functions in a clinical setting for diagnostic and prognostic purposes and eventually in clinical trials. Moreover, scoring the performance of such a VSST may have implications in the rehabilitation of cognitive functions and in general may be used for upgrading mental activity by exercise.

In the vast majority of the literature on eye movements, saccade amplitude or duration, number of fixations, fixation durations, or other close derivatives have been used as the main measures (see for instance the recent contribution in [7]). Although saccades and fixations are fundamentally sequential, very few methods are available for treating their sequential properties. Among those taking into account the fixation order, the most widely applied method is based on the edit distance, i.e., minimum number of "edit" operations transforming a sequence into another [8]. More advanced versions assign different weights to each operation. Such methods have been successfully used by a number of researchers to study saccade sequences (e.g., [9–13]). These methods define a number of spatial ROIs in the scene being scanned and the fixation sequence is coded as a series of letters representing the fixated locations. Although the string edit method has proven to be a useful tool and is relatively fast to compute, one of its main drawbacks is that it does not take the relationship between ROIs into consideration, so that the algorithm cannot differentiate between close and distant ROIs. A second drawback of this kind of method is that they do not take the fixation duration into account; all fixations, however short or long, are treated equally. Instead, it is clear that the fixation duration is an important indicator of processing during a fixation [14]. In [15], Cristino et al. describe a new method, for quantitatively scoring two eye movement sequences: they show how the methodology of global sequence alignment (Needleman–Wunsch algorithm [16]) can be applied to eye movements and then present three experiments in which the method is used.

In this paper, we follow the approach to take the fixation order, the fixation duration and the spatial distance from the ROIs into account. First, we pre-process the data recording the fixation sequence as a series of symbols (possible repeated) representing the fixated locations. Since the observed sequences (scan-paths) have a length quite different from each other, a global alignment is not suitable to evidence their similarity (if any) [17]. Therefore, we at first propose to compare the expected scan-path with the observed scan-path using dot-plots. This provides a visual and hence a qualitative comparison between them but does not permit to evaluate it quantitatively.

From a different point of view, the problem we want to tackle is also related to the so-called episode matching [18]. An episode is a collection of events that occur within a short time interval. In our case, an event corresponds to a fixation, and an episode to a scan-path. Usually, in the episode matching problem, given a long sequence of events, it can be useful to know what episodes occur frequently in the sequence.

A simplified version of this problem can be restated as an approximate string matching problem [19]: Given a text T, find its substrings containing the string P as a subsequence. Conditions on the number of occurrences and/or on the length of the substrings of T can be considered. Here, we investigate the problem in which T is the obtained scan-path, and P is the task scan-path.

In particular, for every obtained scan-path, we determine the first occurrence of P in T, and we score it. A novel scoring scheme is presented that takes into account the spatial relationship between ROIs (differentiate between close and far regions—distance matrix) and the fixation duration (repetitions of the letter corresponding to the ROI in a way that is proportional to the fixation). It also includes the guess that fixations outside the ROIs may be part of the exploration strategy.

The proposed score is validated by comparing the performance of the three different groups: the group of patients with extrapyramidal disease, the second one of patients suffering from chronic pain syndrome and the control group. Our results, as expected, confirm the worst performance of extrapyramidal patients than the chronic pain and control groups, in general. In particular, the medians of the three classes are significantly different from each other, so suggesting that our method can be employed as a measure of the performance in the VSST.

Summarizing, the main contributions of this paper are:

- A new way to preprocess the VSST data, so as to represent them as sequences to which classical alignment methods can be applied;
- A novel scoring scheme to evaluate the observed scan-path with respect to the target scan–path;
- A preliminary experimental analysis on an original VSST dataset which highlights different pathological behaviours validated by human experts.

The method we propose is illustrated in the flowchart of Figure 1. The paper is organized as follows. In the next section, the task that we want to pursue is described, together with the data pre-processing and the proposed alignment approach, based on a new ad hoc definition of the similarity score. Section 3 collects experimental results that are discussed in the following Section 4. Finally, in Section 5, some conclusions are drawn and also open questions and future perspectives are described.

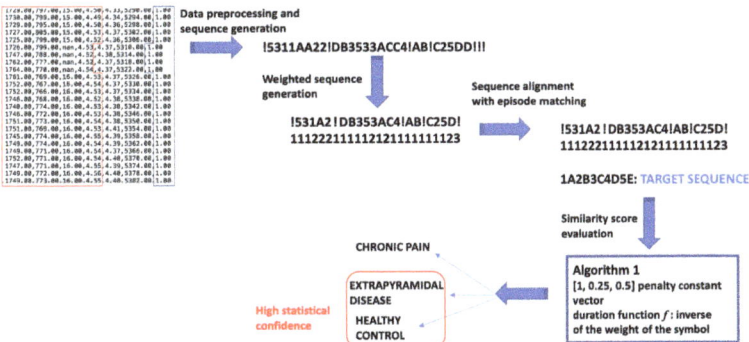

Figure 1. Processing pipeline: average gaze position per timestamp in fixation rows are processed to get the scan-path observed by the patient (the columns related to the pupils' size are not taken into account); after creating the weight vector, we run the similarity score (with respect to the target sequence) to get the final patient score. Healthy patients and patients with extrapyramidal nervous system disease are distinguished with high statistical confidence.

2. Material and Methods

2.1. Task Design

There exists several different TMT settings that can be adopted. For instance, a patient could be supposed to link ordered series of numbers or letters (which we will generally call *symbols* in the following paragraphs) drawing with paper and pencil [20] or onto an electronic device [21,22]. In other settings, tested people are required to sit in front of a monitor and interact with screen-based content, through an eye-tracker device [23,24]. Our study is carried out based on this last setting, which allows us to perform a Visual Sequential Search Test (VSST). In particular, the stimulus images submitted (in this order) to the patient are illustrated in Figure 2, and the required task is to make the sequence 1-A-2-B-3-C-4-D-5-E at least once during the whole test time.

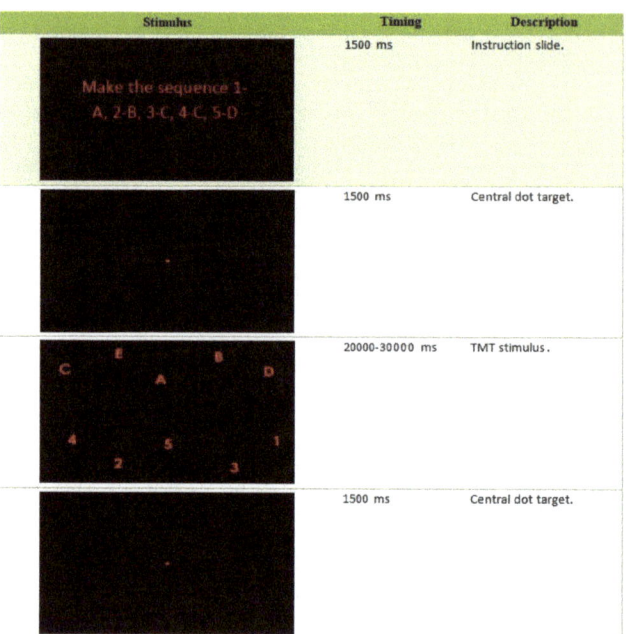

Figure 2. Stimuli timing: the instruction slide is codified with "NaN"; the central dot target is codified with "0"; the TMT stimulus is codified with "1".

2.2. Data Preprocessing

2.2.1. Dataset

The data obtained by the eye-tracking experiments, for each person, provide the following information:

- average gaze position (x) (pixels)
- average gaze position (y) (pixels)
- fixation ID (integer) (NaN = saccade)
- pupil size (left eye)
- pupil size (right eye)
- timestamp (every 4 milliseconds)
- stimulus (code of the image shown in the screen): the coding is described in the caption of Figure 2.

Regular eye movements alternate between saccades and visual fixations. A fixation is the maintaining of the visual gaze on a single location. A saccade is a quick, simultaneous movement of both eyes between what happens among two or more phases of fixation in the same direction.

In case of blinking, the device loses the signal and it results in "NaN" (Not a Number) values either for the position (x, y) on the screen and for the pupil sizes. Pupil sizes were not taken into account for the data processing described in the following.

The results of the eye-tracking experiments for 376 subjects were divided into three classes: 46 patients with extrapyramidal syndrome, 284 affected by chronic pain and 46 controls.

It is worth noting that the collected dataset is significantly unbalanced, a problem naturally attributable to the type of pathologies to be prognosed. In particular:

- For extrapyramidal patients the diagnosis is based on objectivable clinical factors (i.e., a movement disorder), while a disability scale exists on which the severity of the disease can be objectively established;
- Chronic pain represents a very variable pathology whose prognosis deeply depends on the personal judgement of human experts and that cannot objectified except through a subjective evaluation scale.

Therefore, an alteration in the scan-path for an extrapyramidal patient is invariably pathological, while a similar alteration evidenced in a patient affected by chronic pain must be treated with caution.

2.2.2. Generated Scan-Path Sequences

Starting from the data previously shown, we dealt with the generation of the scan-path sequences as follows.

The goal here is to use the information of the data to reconstruct the scan-path of an individual during the test as a sequence of symbols, associating a letter or a number for fixations on the ROIs accordingly, and the special character "!" for fixations outside the ROIs (black area). In other words, we generated a string $T = t_1 \ldots t_n$ over the alphabet $\mathcal{A} = \{1, 2, 3, 4, 5, A, B, C, D, E, !\}$. After having determined the centroids of each symbol in the TMT stimulus image, we have calculated the minimum distance between any pair of centroids, and we set a threshold equal to its half. Then, for every fixation ID, we computed the distance from the fixation area to the closer centroid, and we selected it as the associated symbol if the distance was less than the threshold, or "!" otherwise.

For instance, a generated sequence can have the following form:

$$!5311AA22!DB3533ACC4!AB!C25DD!!!...$$

Finally, we stacked subsequent repetitions of symbols in a vector of "weights", associated to the non redundant sequence. Formally speaking, in a string $t_1 \ldots t_n$ where $\exists\, i$ s.t. $t_i = t_{i+1} = \cdots = t_{i+k}$, we replace $t_i, t_{i+1}, \ldots, t_{i+k}$ with $t_{i'} = t_i$ associating the corresponding weight $w_{i'} = k + 1$, with $1 \leq i' \leq n$. This can be easily done by scanning the string and counting the number of consecutive occurrences of the same symbol, in linear time with respect to the length of the string.

For the above sequence, we obtain as the result of the preprocessing of the data:

$$!531A2!DB353AC4!AB!C25D!$$

with the corresponding vector of weights:

$$(1,1,1,2,2,2,1,1,1,1,1,2,1,2,1,1,1,1,1,1,1,2,3)$$

2.3. VSST Data Analysis Method

We are going to formulate the VSST problem in terms of a pairwise sequence alignment, where both the target scan-path and the obtained scan-path are strings.

Let $T = t_1 \ldots t_n$ be any string of length n over the alphabet $\mathcal{A} = \{1, 2, 3, 4, 5, A, B, C, D, E, !\}$, and let $P = 1A2B3C4D5E$. Given T and P, we look for the matches of P in T, that is the occurrences of symbols of P in T. Regions of identity (matches) can be visualized by the so-called dot-plot. A dot-plot is a $10 \times n$ binary matrix M such that the entry $m_{ij} = 1$ if and only if $p_i = t_j$, otherwise $m_{ij} = 0$. Some toy examples are shown in Figure 3 where the identity is visualized by a dot.

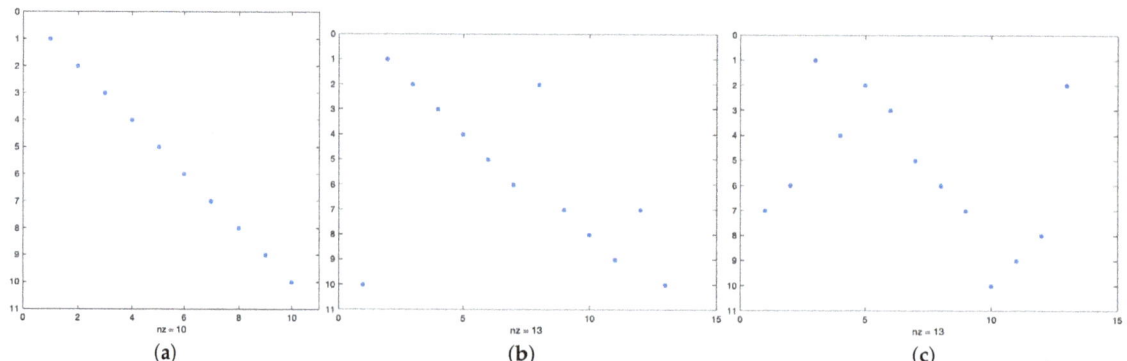

Figure 3. Dot-plots for the toy-sequences: (**a**) T = 1A2B3C4D5E; (**b**) T = E1!A2B3CA4D54E and (**c**) T = 4C1BA2!3C4E5DA.

It is easy to see that "diagonals" of dots correspond to consecutive matches of P in T. This can be formalized as follows. A substring of T is a finite sequence of consecutive symbols of T, while in a subsequence symbols are not necessarily consecutive. Thus, P is a subsequence of T if there exist indices $i_1 < \ldots < i_m$ such that $p_1 = t_{i_1}$, $p_2 = t_{i_2}$, $p_m = t_{i_m}$ and $T' = t_{i_1} t_{i_1+1} \ldots t_{i_m}$ is the substring of T containing P.

Let us define the VSST problem as an approximate string matching problem. The approximate string matching problem looks for those substrings of the text T that can be transformed into pattern P with at most h edit operations: a deletion of a symbol x of T changes the substring uxv into uv; an insertion of a symbol x changes the substring uv of T into uxv; a substitution of a symbol x of T with a symbol y changes the substring uxv into uyv. When deletion is the only edit operation allowed and we choose $h = k - m$, the problem is equivalent to finding all substrings of T of length at most k that contain P of length m as a subsequence.

In the VSST problem we search for the first occurrence of P in T, i.e., we find the substring of T starting in the leftmost symbol in T containing P as a subsequence. This can be done in linear time in the size n of T with the naïve algorithm.

2.4. The Score Scheme

Let $T' = t'_1 \ldots t'_k$ be the substring of T containing P. Next step consists of scoring the approximate matching between T' and P. Actually, $h = k - 10$ provides a first evaluation of the distance between T' and P since they differ by h symbols. Note that this corresponds to defining a scoring system that assigns value 1 to each deletion and sums up each value. However, this measure is oversimple to provide a meaningful evaluation, and moreover we prefer to measure the complementary information, to calculate a "similarity score" between T' and P. Indeed our goal is to assign a final score assessing the performance of the patient in the VSS test. The first step in the definition of the scoring function is to assign a positive value (a reward) to each match, i.e., to each occurrence of a symbol of P in T'. On the contrary, each deletion of symbols of T' must be assigned a negative value (a penalty). We decided to weakly penalize a deletion of the symbol ! with respect to the deletion of any other symbol, since we consider a fixation of the background as an intermediate pause in the process, but not a true selection of an ROI. We refer to these three values as *penalty scale constants*.

In addition, in the latter case (deletion of a symbol not ! in T'), we compute the distance of the centroid of the ROI corresponding to the deleted symbol to the centroid of the ROI of the next expected symbol of P, to take the spatial relation between the two ROIs. The set of the distances for each pair, normalized by the maximum distance, is then collected in a *distance matrix* (Figure 4).

	1	A	2	B	3	C	4	D	5	E
1	0.000000	0.584612	0.710496	0.455064	0.222911	1.000000	0.911789	0.311502	0.484323	0.803034
A	0.584612	0.000000	0.416762	0.279239	0.513616	0.432472	0.476242	0.482918	0.288878	0.219080
2	0.710496	0.416762	0.000000	0.651854	0.516600	0.487140	0.233886	0.783401	0.235490	0.492174
B	0.455064	0.279239	0.651854	0.000000	0.501183	0.691448	0.751307	0.231357	0.450730	0.443744
3	0.222911	0.513616	0.516600	0.501183	0.000000	0.879225	0.735605	0.456362	0.315470	0.725935
C	1.000000	0.432472	0.487140	0.691448	0.879225	0.000000	0.317676	0.911267	0.571901	0.255773
4	0.911789	0.476242	0.233886	0.751307	0.735605	0.317676	0.000000	0.927189	0.428144	0.439809
D	0.311502	0.482918	0.783401	0.231357	0.456362	0.911267	0.927189	0.000000	0.553153	0.670334
5	0.484323	0.288878	0.235490	0.450730	0.315470	0.571901	0.428144	0.553153	0.000000	0.460523
E	0.803034	0.219080	0.492174	0.443744	0.725935	0.255773	0.439809	0.670334	0.460523	0.000000

Figure 4. Distance matrix.

Another factor included in the score is the duration of the fixation. We store the information in a parallel array as explained in Section 2.2.2. We assume that the fixation duration is associated to hesitation in the VSST. Since duration corresponds to consecutive repetitions of any symbol, we define a function decreasing in the number of repetitions for scoring the match and increasing in the number of repetitions for scoring the deletion. We refer to it as the *duration function*.

Finally, since the fixations outside the ROIs may be part of the exploration strategy, we compute the frequency of each symbol in the prefix ending there, to amplify the penalty: the frequency corresponds to the number of times that the symbol has been already fixed in the exploration so that it reflects the number of times needed to learn its position.

To summarize, the final score of T' is the sum of the contributions to the score for each symbol in T' where each score is obtained by the product of the following factors: the penalty scale constant v, the duration function f, and, in case of deletion of a symbol non !, an item of the distance matrix, $dist$, and the frequency $freq$ of the symbol. The computation of the score is sketched in Algorithm 1.

Algorithm 1 Similarity score evaluation

Require: $T', w, align, v, P, f(w)$
Ensure: $score$
   $j \leftarrow 0$ ▷ index for P
   $i \leftarrow 0$ ▷ index for T'
   $score \leftarrow 0$
   $freq(k) \leftarrow 0 \ \forall k \ in \ P$
   while $j \neq length(P) \ AND \ i \neq length(T')$ **do**
      if $i = align(j)$ **then**   ▷ match
         $p\_score \leftarrow v(0) \cdot f(w(i))$
         $freq(P(j)) \leftarrow freq(P(j)) + 1$
         $j \leftarrow j + 1$
      else if $T'(i) =!$ **then**   ▷ deletion
         $p\_score \leftarrow -v(1) \cdot [1.1 - f(w(i))]$
      else
         $freq(T'(i)) \leftarrow freq(T'(i)) + 1$
         $p\_score \leftarrow -v(2) \cdot freq(T'(i)) \cdot dist(T'(i), P(j)) \cdot [1.1 - f(w(i))]$
      end if
      $score \leftarrow score + p\_score$
      $i \leftarrow i + 1$
   end while

We remark that this algorithm uses three vectors: the substring T', the vector w of the weights of size k and a vector *align* of size $m = 10$, which stores the indices of the items of P such that $align(j) = i$ iff $t_i = p_j$, else $align(j) = -1$. The algorithm scans T' based on the index i and P based on j. Initially $i = j = 0$. Then, it checks if i is equal to $align(j)$: if true, it scores the match (t_i is equal to p_j) and both indices are increased, otherwise it scores the deletion of t_i and then increases i. In case of deletion, it checks if t_i is equal to ! and, consequently, computes the appropriate score. Each access to the vectors takes $O(1)$ and the algorithm scans the whole vector T' so that it runs in $O(k)$ time.

3. Experimental Results

After the pre-processing phase described in Section 2.2.2, the data consist of strings with their weights divided into three classes, depending on the individuals performing the test: 46 strings from patients with extrapyramidal syndrome, 284 from patients affected by chronic pain and 46 healthy participants. From now on, we refer to them as the Extrapyramidal (E), the Chronic (C) and the Healthy (H) classes.

For each member of the classes, we computed the score using the algorithm described in Section 2.4. In particular we used $v = [1, 0.25, 0.5]$ for the penalty constant vector, and the inverse of the weight of the symbol for the duration function f.

Figures 5 and 6 illustrate the dot-plots and the scores computed for a member of each class, respectively. We are going to show that these members are good "representatives" of their classes. At a glance, the dot-plots suggest that the first image corresponds to a performance better than the second, which in turn, looks better than the third.

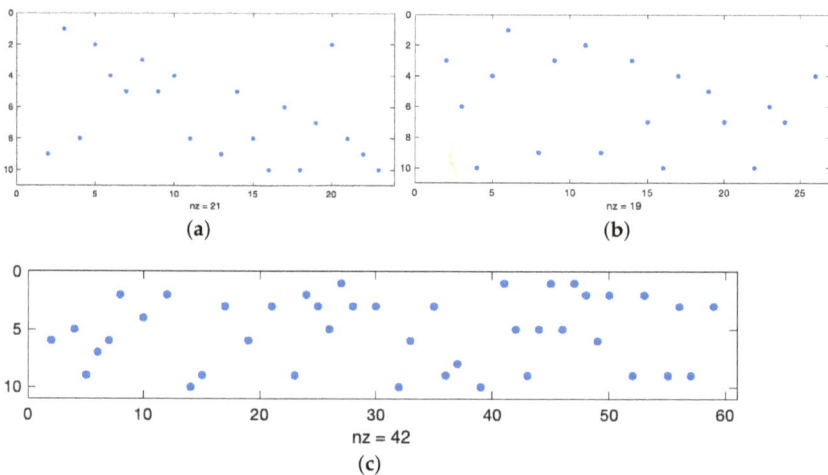

Figure 5. A dot-plot of the strings: (**a**) !5511DAAB3223BBD!53DECCE44ADD55E, member of the Healthy class; (**b**) !!!2CEB1!52!AA55!!24EBBB!!!334!ECC4!B, member of the Chronic class; (**c**) !!!C!3354CCAA!B!A!!E5!!!2!!C!2!5A23122!2!EC!25D!EEE!1353131ACCA!5A!525!2!, member of the Extrapyramidal class.

In the images of Figure 6, we illustrate the score as the bar graph obtained by visualizing each value assigned to each symbol of the sequences as a bar. Let us notice that bars of positive height correspond to the score of matches, whereas bars of negative height correspond to the score of deletions. Matches can be scored with values lower than 1, when repeated; deletions are scored differently depending on repetition, frequency, and distance from the next symbol objective.

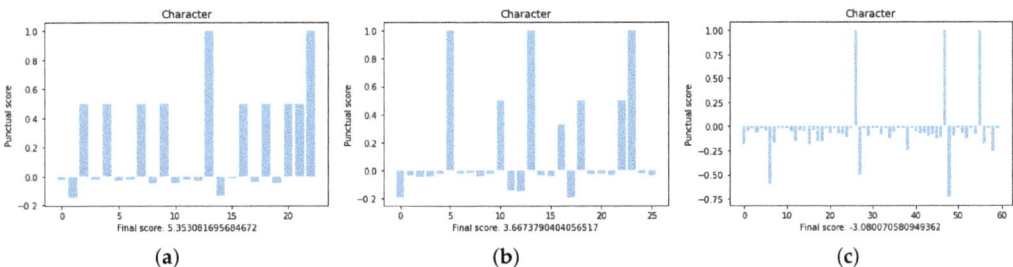

Figure 6. A bar graph of the scores of the strings: (**a**) !5511DAAB3223BBD!53DECCE44ADD55E, member of the Healthy class; (**b**) !!!2CEB1!52!AA55!!24EBBB!!!334!ECC4!B, member of the Chronic class; (**c**) !!!C!3354CCAA!B!A!!E5!!!2!!C!2!5A23122!2!EC!25D!EEE!1353131ACCA!5A!525!2!, member of the Extrapyramidal class.

Before the analysis, we dropped some outliers for each class, according to the Chebischev Theorem. Setting $\gamma = 2$, we were sure to retain at least 75% for each class; such a dropping resulted in retaining 43 sequences out of 46 for the Healthy class, 265 out of 284 for the Chronic class and 44 out of 46 for the Extrapyramidal class.

Therefore, we analysed the results using the R language for computing the basic statistics and graphics. A summary divided by the groups is shown in Figure 7, while in Figure 8 we report the box plots.

	Healthy	Chronic	Extrapyramidal
Min	-11.1419	-14.2344	-16.6293
1st quartile	2.8705	0.3399	-7.0600
Median	5.3861	4.1925	-2.8794
Mean	3.8202	2.1650	-2.7150
3rd quartile	6.5742	5.9017	0.8144
Max	8.3623	9.7753	8.4175

Figure 7. Main statistics for the three classes of patients.

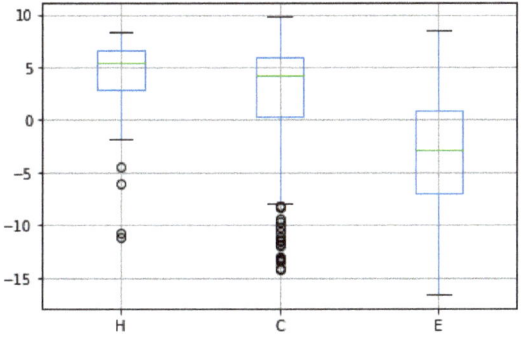

Figure 8. Box plots of the score distribution for each class: Healthy (**left**), Chronic (**center**), and Extrapyramidal (**right**).

Based on the obtained results, we can notice that the data seems not to follow a normal distribution, as we can see from Figure 9, at least for two of the three classes (the Healthy class and the Chronic one). Indeed we run the Kolgomorov–Smirnov test for comparison with the normal distribution on each class, obtaining p-values, respectively, equal to $6.56388017310154 \cdot 10^{-30}$, $9.711223032427313 \cdot 10^{-105}$, $1.5674138951676932 \cdot 10^{-12}$.

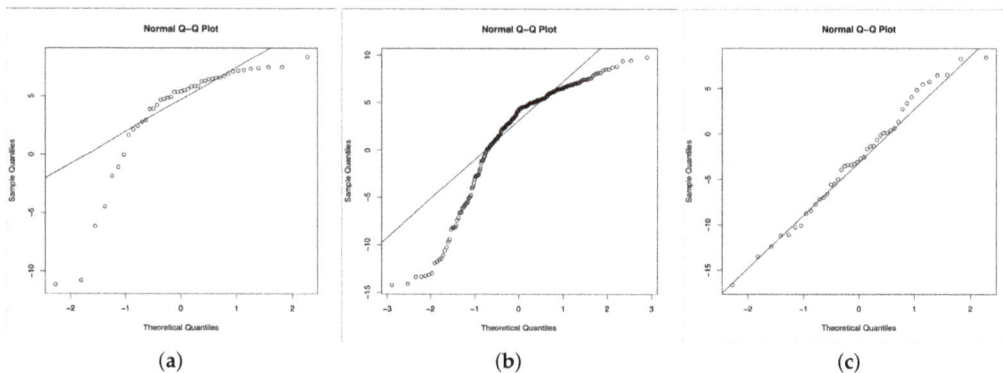

Figure 9. Q-Q plots for each patients' class: (**a**) Healthy, (**b**) Chronic, and (**c**) Extrapyramidal.

Thus, we used the non-parametric Kruskal–Wallis test by rank which extends the two-sample Wilcoxon test in the situation where there are more than two groups. It turns out that at 0.05 significance level, the medians of the data of the three groups are different. In particular, the p-value for the Kruskal–Wallis test is p-value = $6.553 \cdot 10^{-8}$. In order to know which pairs of groups are significantly different we used the function pairwise.wilcox.test() to calculate pairwise comparisons between group levels with corrections for multiple testing and Bonferroni correction. The results confirm that the pair exhibiting the most significant difference is the Healthy–Extrapyramidal as expected (see Table 1). Indeed, patients with extrapyramidal disease have well known difficulties in visual spatial exploration and executive functions that result in difficulties from the subject to maintain a top-down (human intention) internal representation of the visual scene during task execution. This is reflected in a bad performance in the VSST. Differently, patients in the Chronic class are affected by several kinds of chronic pain syndromes so that they may have different behaviours in performing the task.

Nevertheless, note that, actually, all the pairs have p-values less than 0.05 so that they are significantly different.

Table 1. p-values of pairwise comparisons using Wilcoxon rank sum test with continuity correction.

	Chronic	Extrapyramidal
Extrapyramidal	$2 \cdot 10^{-6}$	-
Healthy	0.04	$8 \cdot 10^{-7}$

4. Discussion

In this study, we propose a method for the analysis of gaze in a top-down visual search task and find a score for the VSST performance. The whole pipeline for the process is illustrated in Figure 1. The considered method and score have been validated by comparing the performance of three different subjects' groups. The first group includes 46 patients with extrapyramidal disease, who have well known difficulties in visual spatial exploration and executive functions. The second group is composed of 238 patients suffering from chronic pain syndrome and the third, collecting 46 patients, is a control group.

The identification of a robust method of analysis and the detection of a reliable indicator for the VSST performance, would allow to give a measure of executive functions in a clinical setting for diagnostic and prognostic purposes and eventually in clinical trials. Moreover, scoring the performance of such a VSST may have implications in the rehabilitation of cognitive functions and in general may be used for upgrading mental activity by exercise.

Indeed, cognitive rehabilitation is an effective non–pharmacological treatment that consists of learning compensatory strategies and exploiting residual skills in order to counteract, for instance, cognitive impairments and degenerative diseases. In fact, as for dementia, unfortunately, there is no specific pharmacological treatment, being existing drugs able to counteract the symptoms of the disease, but do not change its course. Consequently, the disease progresses: there is a continuous and constant progressive decline of cognitive functions for the patient, which negatively affects the various daily skills. Instead, changing the course of the disease, "pushing forward" the degenerative progression allows the patient to maintain their autonomy for a longer time and reduces the disinterest, anxiety and depression that degenerative diseases entail. Finally, cognitive rehabilitation is also fundamental for maintaining cognitive functions in efficiency and to combat the consequences of normal aging. Similarly, it is possible to implement intellectual stimulation with a preventive purpose.

The main characteristic of VSST is that it forces the subject to perform a default and logic path using high level cognitive resources. In this task the target of the next fixation changes continuously and, thus, in order to perform an adequate eye movement, each fixation must contain the information on the current target position and the next target location [25]. Previously, Veneri et al. [26] suggested that the re-sampling of the spatial element in such a visual search task requires a ranking of each element of the sequence during fixations. To be effective, this process requires a maximization of the discrimination abilities of the peripheral vision. The comparison of the *expected scan-path* with *the observed scan-path* provides a valuable method to investigate how a task forces the subject to maintain a top-down (human intention) internal representation of the visual scene during task execution. The proposed method has proved to be really effective in distinguishing between healthy people and patients affected by extrapyramidal pathologies, and less sensitive to the differences among the other cohort combinations. Actually, patients with chronic pain syndrome may be affected by very different pathologies—from severe neoplasms to chronic migraines—not all equally disabling from a neurocognitive point of view, which makes this group of patients extremely heterogeneous and difficult to distinguish, for example, from healthy people. Anyway, the main advantage of the proposed VSS test, equipped with the automatic procedure to score its outcomes, lies in the possibility of standardizing the test—making the obtained results repeatable—as well as memorizing them permanently. In this way, for each patient, a historical series of their performance can reliably be collected and analysed, a suitable procedure for evaluating both the course of a disease or the recovery based on a cognitive rehabilitation process.

Some issues concerning the proposed method naturally arise:

- we are aware of the approximation errors that could occur by converting gaze positions into symbols through the ROIs processing, as a threshold has been set to discriminate from region to region. Nevertheless, comparing the retrieved sequences to video analysis, our method seems to catch efficiently the sequence of symbols scanned during the test;
- the proposed score algorithm takes into account some factors, such as the duration of the fixation, or the distance from the target symbol. Surely they are not the only factors linked to cognitive tasks involved in the test: improvements in building up the algorithm could be considered by taking into account other minor cognitive factors;
- the weight function, applied on the stacked subsequent repetitions of single symbols, has been arbitrarily chosen as the inverse function. Future works could focus on

designing a more appropriate weight function, which could take into account cognitive features associated with the repetitive behaviour.

5. Conclusions

In this manuscript, we have described a new method for the analysis of the Visual Sequential Search Test, a neurocognitive task commonly used in clinical settings as a diagnostic tool for the evaluation of frontal functions. The VSS test is an eye-tracking version of the Trail Making Test to discover how selection (fixations) guides next exploration (saccades), and how human top-down factors interact with bottom-up saliency. The problem of analysing the VSST outcomes is faced as an episode–matching problem, where an event corresponds to a fixation, and an episode to a scan–path. In this way, a score can be devised able to quantify how much a particular outcome diverges from the expected one. Based on this score, we are able to predict, with a high statistical confidence, if a particular scan-path corresponds to a patient with an extrapyramidal disease or suffering from the chronic pain syndrome or if it describes a "normal" cognitive behaviour. Having a standardized way to evaluate the VSST can help for monitoring the evolution of a disease, for neurological rehabilitation and for intellectual stimulation with a preventive purpose. In particular, the preventive aspect is taking on an increasingly important role, both in terms of the physical and intellectual well–being of the population, and in a more general process of optimizing economic resources for healthcare. We are perfectly aware of the small size of the used dataset. This is a common problem with medical data. Apart from collecting new data, another possible way to overcome this problem could consist in applying data augmentation techniques in order to both balance and enlarge its size. This will be an issue to discuss in future investigations.

Author Contributions: Conceptualization, G.A.D., S.B. and M.L.S.; Data curation, S.B.; Formal analysis, G.A.D., S.B. and M.L.S.; Investigation, G.A.D., S.B., M.L.S., A.R. and M.B.; Methodology, G.A.D., S.B., A.R. and M.B.; Project administration, M.B.; Resources, D.F.M.; Software, G.A.D., S.B. and M.L.S.; Supervision, M.L.S., A.R. and M.B.; Validation, A.R.; Visualization, G.A.D., S.B. and M.L.S.; Writing — original draft, G.A.D., S.B., M.L.S., A.R. and M.B. All authors have read and agreed to the published version of the manuscript.

Funding: This research received no external funding.

Institutional Review Board Statement: Not applicable.

Informed Consent Statement: Patient consent was waived due to the anonymous nature of analyzed data.

Data Availability Statement: Not applicable.

Acknowledgments: The authors wish to thank RoNeuro Institute, part of the Romanian Foundation for the Study of Nanoneurosciences and Neuroregeneration, Cluj-Napoca, Romania, represented by Dafin Fior Muresanu, for providing the datasets used here for the experiments.

Conflicts of Interest: The authors declare no conflict of interest.

References

1. Reitan, R.; Wolfson, D. *The Halstead-Reitan Neuropsychological Test Battery*; Horton, A.M., Jr., Webster, J.S., Eds.; John Wiley & Sons: Tuscon, AZ, USA, 1985.
2. Veneri, G.; Pretegiani, E.; Rosini, F.; Federighi, P.; Federico, A.; Rufa, A. Evaluating the human ongoing visual search performance by eye tracking application and sequencing tests. *Comput. Methods Programs Biomed.* **2012**, *107*, 468–477. [CrossRef]
3. Hochstadt, J. Set-shifting and the on-line processing of relative clauses in Parkinson's disease: Results from a novel eye-tracking method. *Cortex* **2009**, *45*, 991–1011. [CrossRef] [PubMed]
4. Marx, S.; Respondek, G.; Stamelou, M.; Dowiasch, S.; Stoll, J.; Bremmer, F.; Oertel, W.H.; Höglinger, G.U.; Einhauser, W. Validation of mobile eye-tracking as novel and efficient means for differentiating progressive supranuclear palsy from Parkinson's disease. *Front. Behav. Neurosci.* **2012**, *6*, 88. [CrossRef] [PubMed]
5. Kaufmann, B.C.; Cazzoli, D.; Pflugshaupt, T.; Bohlhalter, S.; Vanbellingen, T.; Müri, R.M.; Nef, T.; Nyffeler, T. Eyetracking during free visual exploration detects neglect more reliably than paper-pencil tests. *Cortex* **2020**, *129*, 223–235. [CrossRef] [PubMed]
6. Trepagnier, C. Tracking gaze of patients with visuospatial neglect. *Top. Stroke Rehabil.* **2002**, *8*, 79–88. [CrossRef] [PubMed]

7. Pancino, N.; Graziani, G.; Lachi, V.; Sampoli, M.; Stefanescu, E.; Bianchini, M.; Dimitri, G.M. A Mixed Statistical/Machine Learning Approach for the Analysis of Multimodal Trail Making Test Data. *Preprint* **2021**. under review.
8. Crochemore, M.; Rytter, W. *Jewels of Stringology*; World Scientific: Hackensack, NJ, USA, 2003.
9. Brandt, S.A.; Stark, L.W. Spontaneous eye movements during visual imagery reflect the content of the visual scene. *J. Cogn. Neurosci.* **1997**, *9*, 27–38. [CrossRef]
10. Choi, Y.S.; Mosley, A.D.; Stark, L.W. String editing analysis of human visual search. *Optom. Vis. Sci.* **1995**, *72*, 439–451. [CrossRef] [PubMed]
11. Foulsham, T.; Underwood, G. What can saliency models predict about eye movements? Spatial and sequential aspects of fixations during encoding and recognition. *J. Vis.* **2008**, *8*, 1–17. [CrossRef]
12. Hacisalihzade, S.S.; Stark, L.W.; Allen, J.S. Visual perception and sequences of eye movement fixations: A stochastic modeling approach. *IEEE Trans. Syst. Man, Cybern.* **1992**, *22*, 474–481. [CrossRef]
13. Noton, D.; Stark, L. Scanpaths in eye movements during pattern perception. *Science* **1971**, *171*, 308–311. [CrossRef] [PubMed]
14. Henderson, J.M.; Pierce, G.L. Eye movements during scene viewing: Evidence for mixed control of fixation durations. *Psychon. Bull. Rev.* **2008**, *15*, 566–573. [CrossRef]
15. Cristino, F.; Mathôt, S.; Theeuwes, J.; Gilchrist, I.D. ScanMatch: A novel method for comparing fixation sequences. *Behav. Res. Methods* **2010**, *42*, 692–700. [CrossRef]
16. Needleman, S.B.; Wunsch, C.D. A general method applicable to the search for similarities in the amino acid sequence of two proteins. *J. Mol. Biol.* **1970**, *48*, 443–453. [CrossRef]
17. Durbin, R.; Eddy, S.R.; Krogh, A.; Mitchison, G. *Biological Sequence Analysis: Probabilistic Models of Proteins and Nucleic Acids*; Cambridge University Press: Cambridge, UK, 1998. [CrossRef]
18. Das, G.; Fleischer, R.; Gasieniec, L.; Gunopulos, D.; Kärkkäinen, J. Episode matching. In *Annual Symposium on Combinatorial Pattern Matching*; Springer: Berlin/Heidelberg, Germany, 1997; pp. 12–27. [CrossRef]
19. Apostolico, A.; Atallah, M.J. Compact recognizers of episode sequences. *Inf. Comput.* **2002**, *174*, 180–192. [CrossRef]
20. Zakzanis, K.K.; Mraz, R.; Graham, S.J. An fMRI study of the trail making test. *Neuropsychologia* **2005**, *43*, 1878–1886. [CrossRef] [PubMed]
21. Bracken, M.R.; Mazur-Mosiewicz, A.; Glazek, K. Trail Making Test: Comparison of paper-and-pencil and electronic versions. *Appl. Neuropsychol. Adult* **2018**, *26*, 522–532. [CrossRef] [PubMed]
22. Drapeau, C.E.; Bastien-Toniazzo, M.; Rous, C.; Carlier, M. Nonequivalence of computerized and paper-and-pencil versions of Trail Making Test. *Percept. Mot. Skills* **2007**, *104*, 785–791. [CrossRef]
23. Hicks, S.L.; Sharma, R.; Khan, A.N.; Berna, C.M.; Waldecker, A.; Talbot, K.; Kennard, C.; Turner, M.R. An eye-tracking version of the trail-making test. *PLoS ONE* **2013**, *8*, e84061. [CrossRef]
24. Jyotsna, C.; Amudha, J.; Rao, R.; Nayar, R. Intelligent gaze tracking approach for trail making test. *J. Intell. Fuzzy Syst.* **2020**, *38*, 6299–6310. [CrossRef]
25. Veneri, G.; Rufa, A. Extrafoveal Vision Maximizes the Likelihood to Grab Information in Visual-sequential Search. In *Computer Communication & Collaboration*; Academic Research Centre of Canada: Ottawa, ON, Canada, 2017; Volume 5.
26. Veneri, G.; Rosini, F.; Federighi, P.; Federico, A.; Rufa, A. Evaluating gaze control on a multi–target sequencing task: The distribution of fixations is evidence of exploration optimization. *Comput. Biol. Med.* **2012**, *42*, 235–244. [CrossRef] [PubMed]

Article

A Mixed Statistical and Machine Learning Approach for the Analysis of Multimodal Trail Making Test Data

Niccolò Pancino [1,2,†], Caterina Graziani [2,†], Veronica Lachi [2,†], Maria Lucia Sampoli [2], Emanuel Ștefănescu [3,4], Monica Bianchini [2] and Giovanna Maria Dimitri [2,5,*]

Citation: Pancino, N.; Graziani, C.; Lachi, V.; Sampoli, M.L.; Ștefănescu, E.; Bianchini, M.; Dimitri, G.M. A Mixed Statistical and Machine Learning Approach for the Analysis of Multimodal Trail Making Test Data. *Mathematics* **2021**, *9*, 3159. https://doi.org/10.3390/math9243159

Academic Editor: Ioannis G. Tsoulos

Received: 18 October 2021
Accepted: 6 December 2021
Published: 8 December 2021

Publisher's Note: MDPI stays neutral with regard to jurisdictional claims in published maps and institutional affiliations.

Copyright: © 2021 by the authors. Licensee MDPI, Basel, Switzerland. This article is an open access article distributed under the terms and conditions of the Creative Commons Attribution (CC BY) license (https://creativecommons.org/licenses/by/4.0/).

[1] Dipartimento di Ingegneria dell'Informazione, Università degli Studi di Firenze, 50121 Firenze, Italy; niccolo.pancino@unifi.it
[2] Dipartimento di Ingegneria dell'Informazione e Scienze Matematiche, Università degli Studi di Siena, 53100 Siena, Italy; caterina.graziani@student.unisi.it (C.G.); veronica.lachi@student.unisi.it (V.L.); marialucia.sampoli@unisi.it (M.L.S.); monica.bianchini@unisi.it (M.B.)
[3] Department of Neurosciences, "Iuliu Hațieganu" University of Medicine and Pharmacy, 400000 Cluj-Napoca, Romania; stefanescu.emanuel@yahoo.com
[4] RoNeuro Institute for Neurological Research and Diagnostic, 400364 Cluj-Napoca, Romania
[5] Dipartimento di Informatica, Università di Pisa, 56127 Pisa, Italy
* Correspondence: giovanna.dimitri@unisi.it
† These authors contributed equally to this work.

Abstract: Eye-tracking can offer a novel clinical practice and a non-invasive tool to detect neuropathological syndromes. In this paper, we show some analysis on data obtained from the visual sequential search test. Indeed, such a test can be used to evaluate the capacity of looking at objects in a specific order, and its successful execution requires the optimization of the perceptual resources of foveal and extrafoveal vision. The main objective of this work is to detect if some patterns can be found within the data, to discern among people with chronic pain, extrapyramidal patients and healthy controls. We employed statistical tests to evaluate differences among groups, considering three novel indicators: blinking rate, average blinking duration and maximum pupil size variation. Additionally, to divide the three patient groups based on scan-path images—which appear very noisy and all similar to each other—we applied deep learning techniques to embed them into a larger transformed space. We then applied a clustering approach to correctly detect and classify the three cohorts. Preliminary experiments show promising results.

Keywords: eye tracking; Til Making Test; visual sequential search test; neurological diseases; deep learning

1. Introduction and Related Work

Eye-tracking offers a fundamental tool to process and analyse human brain behaviour by detecting eye position and speed of movements [1]. Moreover, eye movements could in principle be used in order to highlight the presence of pathological states, and consistent research has been recently performed in this direction [2,3]. In the last decades, Machine Learning (ML) has been widely applied to many different research fields [4–7] and, in particular, some examples of its use for Trail Making Test (TMT) data analysis can be found in the literature. For instance, in [8], an approach based on random forests, decision trees and Long Short Term Memories (LSTMs) was proposed to detect the presence of a pathological state in the tested subjects. In particular, 60 patients were recruited in the study, 24 of which presenting brain injury and 36 presenting vertigo episodes. Similarly, in [9], the eye-tracking test was used to analyse children diagnosed with autism spectrum disorder (ASD), in order to establish a quantitative relationship between their gaze performance and their ability in social communication. Indeed, in the same study, the eye gaze-tracking was proposed as a possible non-invasive, quantitative biomarker to be used in children with ASD. Finally, a vast literature exists related to applications of eye-tracking tests to detect depression syndromes [10–13], and eye-tracking studies have proved their efficacy in the diagnosis of other common neurological pathologies, such as Parkinson's disease, brain trauma and neglect phenomena [14–17], while ML techniques have been recently applied to process TMT data for the detection of the Alzheimer's disease [18,19].

In [20], a new experiment based on TMT has been proposed, called the Visual Sequential Search Test (VSST). In a standard TMT experiment, a subject is presented with a sheet of numbers and letters arranged in a random manner and is asked, using a pen, to perform two tasks simultaneously, namely, to connect both in progressive and alternating order numbers and letters. In the VSST setting, the patients are required to carry on the same task based only on eye movements. Human visual search [20,21] is, in fact, a common activity that enables humans to explore the external environment to take everyday life decisions. Indeed, sequential visual search should use a peripheral spatial scene classification technique to put the next target in the sequence in the correct order, a strategy which, as a byproduct, could also improve the discriminatory ability of human peripheral vision and save neural resources associated with foveation. With respect to the cohorts of patients under examination, data were collected from people with chronic pain, extrapyramidal patients and healthy controls. In particular, individuals affected by extrapyramidal symptoms suffer from tremors, rigidity, spasms, decline in cognitive functions (dementia), affective disorders, depression, amnesia, involuntary and hyperkinetic jerky movements, slowing of voluntary movements such as walking (bradykinesia), and postural abnormalities. Conversely, there are several mechanisms underlying chronic pain; more often an excessive and persistent stimulation of the "nociceptors" or a lesion of the peripheral or central nervous system, but there are also forms of chronic pain that do not seem to have a real, well-identified cause (neuropathic pain). Therefore, chronic pain can be related to a variety of diseases, with very different severity, from depression, to chronic migraine and to cancer.

In [22], an algorithmic approach for the analysis of the VSST, based on the episode matching method, is proposed. In this paper, instead, we analyse the VSST data from a different perspective, examining both the blinking behaviour and the pupil size of the subjects and the freezed images of the scan-paths captured during the test, to gain an insight into the patient condition and offer a support for the clinical practice. For this purpose, we have compared several indicators to distinguish among classes of patients. A first preliminary analysis was performed, to evidence statistical differences based on pupil-derived measures. Such analysis showed the presence of statistically diversified behaviours existing among healthy, chronic and extrapyramidal subjects. Moreover, we further implemented a Deep Learning (DL) autoencoder architecture, with a U-Net backbone [23], to reconstruct the trajectory images for the three groups of individuals. Subsequently, as a proof of concept, we analysed the latent embedding representations using the K-means clustering algorithm, to verify the presence of clusters corresponding to the three cohorts of patients. Preliminary experiments actually evidence well-defined phenotypical groups in the latent space.

The paper is organised as follows. In Section 2, the VSST and the dataset used are described, together with the statistical methodologies and the DL approach employed for analysing pupils and image data. In Section 3, we summarise and discuss the obtained results. Finally, Section 4 collects some conclusions and traces future work perspectives.

2. Materials & Methods
2.1. Visual Sequential Search Test

The Trail Making Test is used in clinical practice as a neuropsychological assessment of visual attention and task switching. The test investigates the subject's attentive abilities and the capability to quickly switch from a numerical to an alphabetical visual stimulus. Successful performance of the TMT requires a variety of mental abilities, including letter and number recognition, mental flexibility, visual scanning and motor function [24].

The research described in this paper used an oculomotor-based methodology, called eye-tracking, to study cognitive impairments in patients affected by chronic pain and extrapyramidal syndrome. Eye-tracking is in fact a promising way to carry out this kind of cognitive tests, allowing the recording of eye movements, to determine where a person is looking, what the person is looking at and for how long the gaze remains in a particular spot. More precisely, an eye-tracker uses invisible near-infrared light and high-definition cameras to project the light into the eye and record the direction in which it is reflected by the cornea [25]. Advanced algorithms are then used to calculate the position of the eye and determine exactly where it is in focus. This makes it possible to measure the visual behaviour and fine eye movements and allows for a more subtle exploration of cognitive dysfunction in a range of neurological disorders.

Several different eye-tracking devices exist, for example, the screen-based eye-tracker [26]. This type of test requires respondents to sit in front of a monitor and to interact with a screen-based content. In the experiments described in this paper, we made use of a special type of TMT experiment, namely the Visual Sequential Search Test. Such test has been created for studying the top-down visual search, which can be summarised as a series of saccades and fixations. In particular, the VSST consists of a repeated search task, and patients are asked to make the connection by looking at a logical sequence of numbers and letters. Here, the required task is to follow with the movement of the eyes the alphanumeric sequence 1-A, 2-B, 3-C, 4-D, 5-E, as shown in Figure 1.

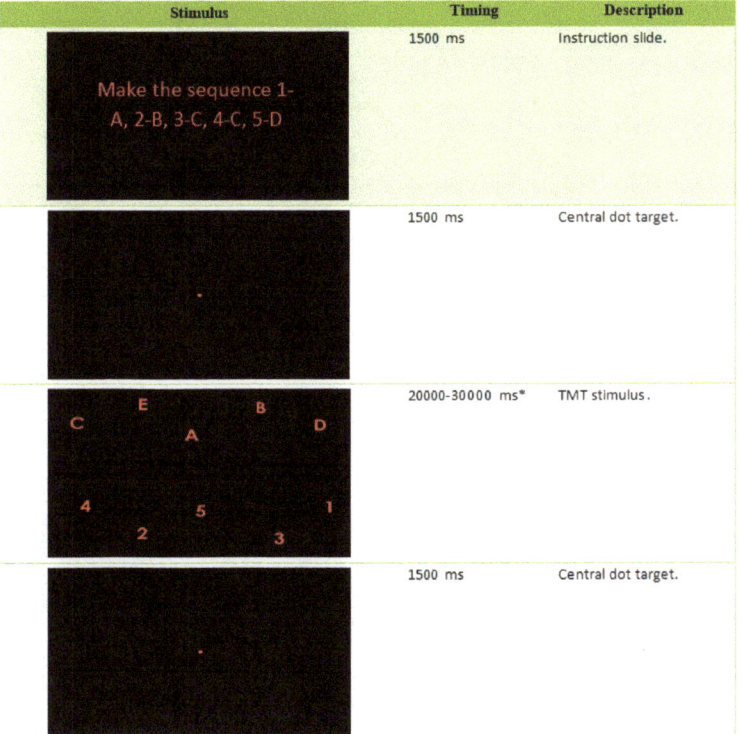

Figure 1. Stimuli timing. * The timing depends on the type of patient to be tested.

2.2. VSST Experimental Dataset Description

Three types of individuals were recruited for the experiments. In particular: 46 patients with extrapyramidal syndrome, 284 affected by chronic pain and 46 controls. For each person, the eye-tracker provided the following information:

- average gaze position along x axis (pixels);
- average gaze position along y axis (pixels);
- fixation ID (integer) (-1 = saccade);
- pupil size (left);
- pupil size (right);
- timestamp (every 4 ms);
- stimulus (code of the image shown in the screen).

Regular eye movements alternate between saccades and visual fixations. A fixation consists in maintaining the visual gaze on a single location. A saccade, instead, is a quick, simultaneous movement of both eyes, between two or more phases of fixation in the same direction. In case of blinking, the device loses the signal, which results in NaN value recorded in our dataset, either for the position (x, y) on the screen and for the size of the pupils.

Data preprocessing was necessary before proceeding with the analysis. In particular, we deleted the experiment part not referring to the image labelled as "TMT stimulus" (Figure 1), we uniformed the timing to have timestamps exactly every 4 ms, and all the artefacts and noisy information were removed from the dataset (e.g., repeated rows).

2.3. ETT Image Dataset

To generate the 2D images of the gaze trajectories, the size of the left pupil and the average positions of the gaze along the horizontal and vertical axes were extracted from the preprocessed numerical data acquired by the eye-tracker during the experiments. In this context, pupil size values equal to NaN correspond to the eye blinks and to those movements recorded while the eye was closed. Therefore these data were removed from the trajectories, as shown in Figure 2.

The ETT (*Eye-Tracking Trajectory*) dataset is composed of images of dimension 1920×1080 pixels, composed by a single colour channel. Binary images constituting the dataset consists of a black background (value 0) where the pixels corresponding to the gaze trajectory appear in white (value 255). In other words, each image is a binarised single-channel (greyscale) image, with intensity values belonging to $\{0, 255\}$. No smoothing operation was performed on the trajectories, in order to preserve the original data information. ETT includes 376 images (46 healthy controls, 46 extrapyramidal patients and 284 chronic pain patients). To generate the dataset, we made use of MATLAB 2021a software [27].

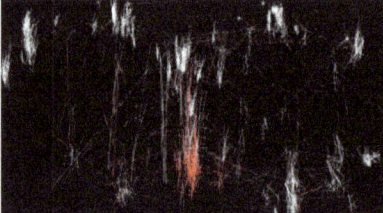 

TMT Trajectories with blinking (in red)   TMT Trajectories with no blinking

Figure 2. An example of a gaze trajectory image of a chronic patient. On the left, the trajectory with blinking movements. Blinking movements are highlighted in red. On the right, the 2D trajectory where the blinking movements are removed.

2.4. Statistical and Deep Learning Methods for the VSST Data Analysis

In the following subsections, we describe how the two sources of information from the VSST can be processed. On the one hand, the behaviour of the population of our patients is analysed with statistical methods applied to the morphological characteristics of the pupil and to the blinking frequency. On the other hand, the images belonging to the ETT dataset were preprocessed on the basis of a DL method, to obtain a latent representation that allows us to adequately group the three cohorts of examined individuals.

2.4.1. Statistical Methods

The following markers were extracted for each patient: the difference between the maximum and the minimum value of the pupil size (averaged over the right and left eye), the blinking rate (i.e., the number of blinks per second) and the blinking average duration. For each of these continuous variables, its distribution over the three classes of patients was computed. Afterwards, a Kruskall–Wallis test [28] was performed. This nonparametric test is used for verifying whether samples originate from the same population (or from populations with equal median). The test has been extensively used for several statistical applications, and has been proved to be a very powerful alternative to parametric tests [29]. It can compare two or more independent samples of different size, testing the null hypothesis H_0, defined by

$$H_0 : \lambda_1 = \lambda_2 = \ldots = \lambda_n,$$

where λ_i is the median of the ith distribution sample.

In order to detect and remove the outliers of each distribution, we applied the unimodal Chebyshev Theorem, with $\gamma = 3$ [30]. This resulted in keeping at least 89% of the values around the mean. Then, we repeated the statistical analysis with the cleaned samples. Finally, a bootstrapping method was performed to avoid the potential effect on the results, due to the difference between the sample size of the chronic pain patients compared to the size of the other two.

2.4.2. Deep Learning Modelling: Autoencoders and the U-Net Architecture

Deep learning has reached state of the art performance in image processing and analysis for a wide range of applications, in particular it gives excellent results in tasks such as image classification, detection or segmentation, see [31–33]. In the present work, we made use of a U-Net-based architecture, a DL model well known for its very good performance in image processing tasks [34]. This model was originally proposed as an efficient and fast way to perform biomedical image segmentation [34]. The architecture is composed by several convolutional layers, which take the original image as input and produce their segmentation maps. It is based on an encoder–decoder structure and can, therefore, be successfully used also to perform image reconstruction. As a matter of fact, in our paper, we trained the U-Net based architecture to reproduce the original image at the output, obtaining a network which is capable of reconstructing the input images (see Figure 3). The architecture used for the present work can thus be viewed as a deep learning, self-supervised autoencoder, made of a downsampling stage (encoder) and an upsampling stage (decoder). The overall scheme of the deep learning architecture proposed, is depicted in Figure 4.

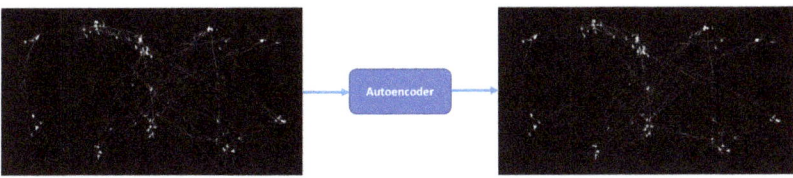

Figure 3. Schematic view of the pipeline implemented for the trajectory image reconstruction.

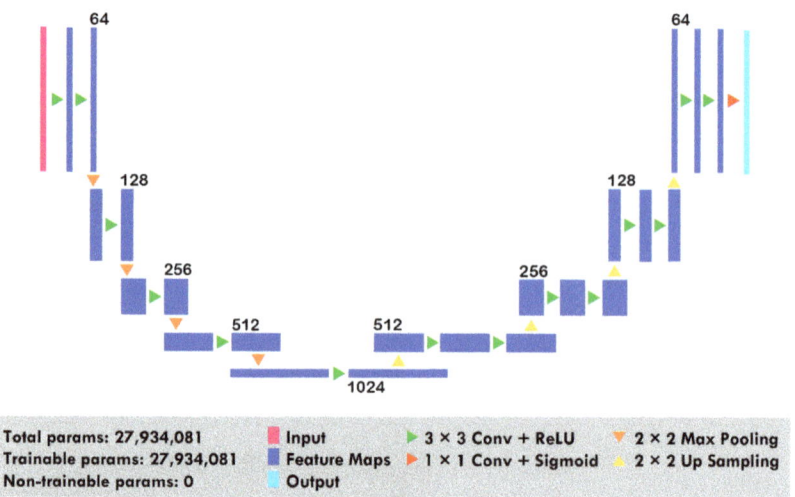

Figure 4. U-Net-based autoencoder architecture used in the experiments. Before both the last layer of the encoder stage and the input layer of the decoder stage, two dropout layers with rate = 0.5 has been used, to force the model not to learn identity function and to prevent overfitting during the training procedure.

In the encoder stage of the network, the spatial dimension is reduced by convolutional blocks followed by a maxpool downsampling layer, while the channel dimension is increased, to encode the input image into a hidden representation at multiple different levels, by means of a series of convolutional layers which use filters to get the so-called *feature maps*. A single feature map provides an insight into the internal representation for the specific input for each of the convolutional layers in the model, capturing some specific information from the input data, such as curves, straight lines or a combination of them. The decoder stage, instead, increases the latent spatial dimension while reducing the number of channels, using convolution blocks followed by an upsampling layer. Generally, the U-Net architecture implies a series of concatenation operations between the output of a layer of the encoder and the input of the corresponding layer in the decoder, by means of residual connections. As the model used acts as an autoencoder, the residual connections have been eliminated, so that the decoder can use only the output of the encoding stage, without including in the reconstruction phase the additional information given by this type of connections. This, also, allows us to avoid overfitting of the reconstruction network, given the small amount of images available. More specifically, 1024 feature maps, each of size 67 × 120 pixels, are obtained from a binarised image of shape 1920 × 1080 × 1, as shown in Figure 5. As a single feature map captures certain aspects from the input data, all the aforementioned 1024 feature maps have been therefore flattened and concatenated to obtain the image embedding representation of shape 1× 8,232,960. Some examples of intermediate representations obtained for a random image for each class are shown in Figure 6.

The model was developed in Python version 3.9.5 with Tensorflow 2.4.0 (Keras backend), and trained using Adam optimizer with an initial learning rate equal to 10^{-4}. All the experiments were performed on a Linux–based machine equipped with an Intel Core i9-10920X CPU, 128 GB DDR4 RAM and a Titan RTX GPU with 24 GB GDDR6 VRAM.

Figure 5. Examples of feature maps from the encoder output for a single input image.

Figure 6. Examples of intermediate representations for three images of healthy controls, extrapyramidal and chronic pain patients, respectively. Each of them has a shape of 1024 × 8040, which is obtained stacking 1024 feature maps composed by 67 × 120 pixels. These latent representations are further flattened to obtain the final embedding of dimension 1 × 8,232,960.

2.4.3. *K*-Means Clustering

As a proof of concept, we performed clustering in both the original and latent space, obtained with our U-Net based model. This would in fact allows to show the ability of the image reconstruction architecture to efficiently compress and maintain the information contained in the original images, producing latent representations which are easier to be distinguished in the three different cohorts [35]. With this intent, we used the *K*-means [36] algorithm, one of the best known and most used partition clustering methods. The algorithm is based on an optimisation process whose aim is to minimise the intra-cluster variance. The number of clusters, *K*, needs to be specified in advance. On the first iteration, *K* clusters are created. Thereafter, the representatives for each cluster are calculated iteratively, until convergence. We used the Scikit-learn Python (Version 3.1).

The *K*-means algorithm has been applied to input data (belonging to the ETT dataset) as well as to the reconstructed data: all the examples in both settings have been grouped in a single cluster, except for only three examples—all belonging to the chronic pain patient category—which have been assigned to the other two clusters. A different strategy was then applied, based on the latent space representations and *K*-means, to determine if collecting the different feature maps, resulting from image compression, was able to lead to a correct grouping of the three categories of patients.

3. Results and Discussion

3.1. Statistical Analysis of Pupil and Blinking Data

First, the Kruskall–Wallis test was applied to the distributions of the blinking rate, maximum pupil size variations and mean blinking duration. In Figure 7, we show the distributions of the three indicators, for healthy, chronic and extrapyramidal individuals. Similarly, in Figure 8, we show boxplots for the three indicators, comparing the three groups.

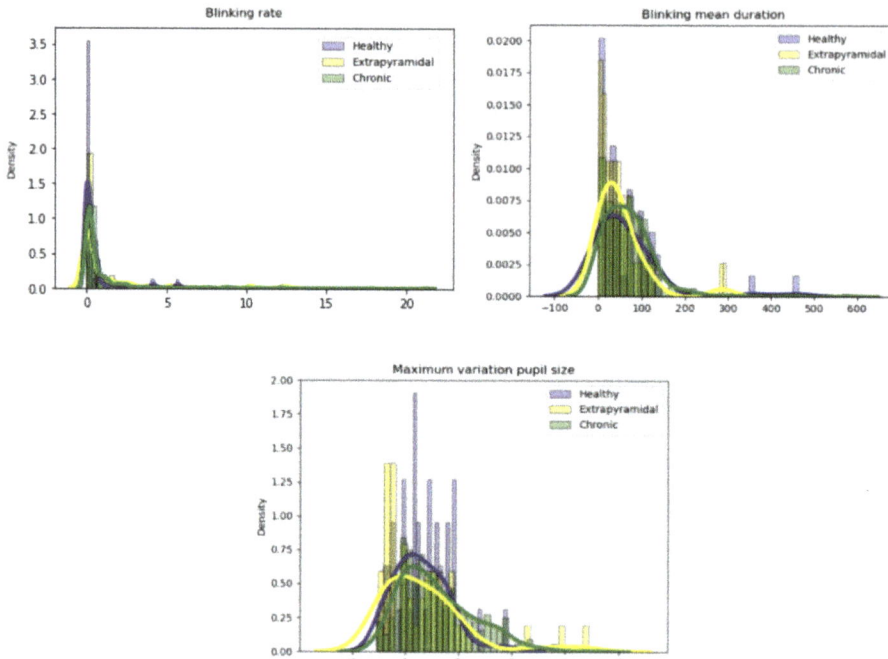

Figure 7. Probability density plots of blinking rate, blinking average duration and maximum variation of the pupil size over the three classes of individuals.

We further performed the Kruskall–Wallis test, to compare the three groups' distributions. The level of statistical significance chosen is p-value = 0.05.

As shown in Table 1, no significant differences are found between healthy subjects and patients affected by extrapyramidal syndrome considering the three indicators. Conversely, a significant statistical difference between healthy controls and chronic pain patients was found for the rate of blinking and the variation of pupil size. Concerning the comparison between patients affected by chronic pain and extrapyramidal syndrome, a significant difference was detected both in the maximum pupil size variation and in the blinking average duration. In Table 1, also the H statistic value is reported, which represents the test statistic for the Kruskal–Wallis test. Under the null hypothesis, the χ-square distribution approximates the distribution of H.

Table 1. Kruskal–Wallis test results of the pairwise comparison between Healthy (H), Chronic (C) and Extrapyramidal (E) subjects for blinking rate, maximum pupil size variation and blinking average duration. Significant p-values are highlighted in bold.

Classes	Blinking Rate		Maximum Pupil Size Variation		Blinking Average Duration	
	p-Value	H Statistic	p-Value	H Statistic	p-Value	H Statistic
H–C	**0.0059**	7.5880	**0.0121**	6.2899	0.1079	2.5847
H–E	0.2092	1.5771	0.6534	0.2016	0.5289	0.3966
E–C	0.3258	0.9654	**0.0039**	8.3394	**0.0058**	7.6226

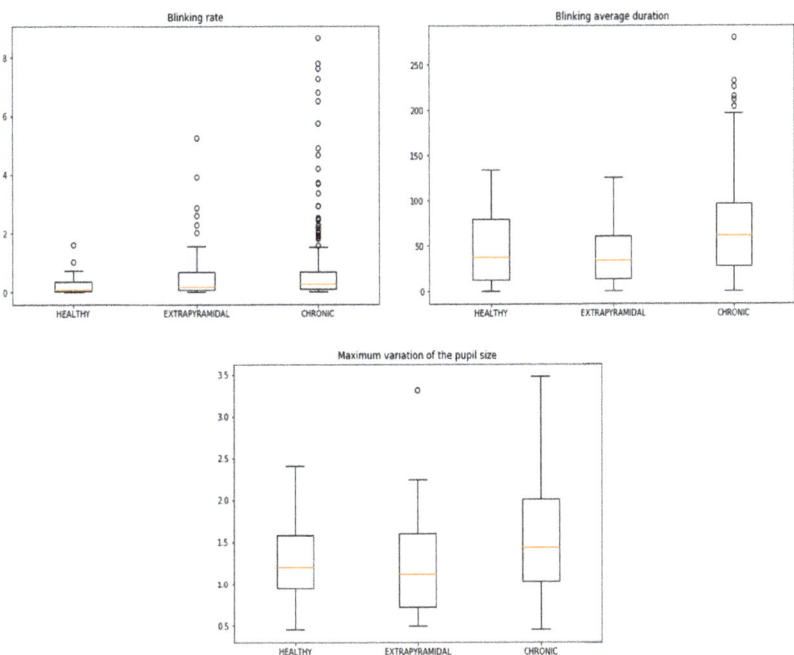

Figure 8. Boxplots of the distribution of the blinking rate, blinking average duration and maximum variation of the pupil size over the three groups. The horizontal line inside the box is located at the median.

3.1.1. Outliers Detection and Kruskall–Wallis Test

A further step of the analysis, as described in Section 2.4.1, consisted in repeating the Kruskal–Wallis test on distributions without outliers. The Chebyshev outlier detection method uses the Chebyshev inequality to calculate the upper and lower outlier detection limits. Data values outside this range will be considered outliers. The outliers could be due to an incorrect acquisition procedure or they could indicate that the data are correct but highly unusual. The results of the Kruskall–Wallis test on the cleaned distributions are reported in Table 2.

The analysis based on clean samples confirmed the previous results: all the significant p-values remained such and, in general, they even decreased. As an effect of this reduction, the difference between Healthy and Chronic subjects in the blinking average duration became significant.

Table 2. Kruskal–Wallis test results of the pairwise comparison between Healthy (H), Chronic (C) and Extrapyramidal (E) subjects for blinking rate, maximum pupil size variation and blinking average duration. Significant p-values are highlighted in bold.

Classes	Blinking Rate		Maximum Pupil Size Variation		Blinking Average Duration	
	p-Value	H Statistic	p-Value	H Statistic	p-Value	H Statistic
H–C	**0.0027**	9.0171	**0.0075**	7.1445	**0.0488**	3.8814
H–E	0.1877	1.7355	0.4984	0.4584	0.5115	0.4309
E–C	0.2390	1.3867	**0.0008**	11.3152	**0.0016**	9.999

3.1.2. Bootstrapping Method

Although the Kruskal–Wallis test is designed for different sample size groups, the greater number of chronic patients than the other two classes may affect the results to some extent. To avoid this kind of bias, we performed the analysis described in the following. A sample of chronic patients was randomly selected from the original distribution, with a size equal to the others—46 patients, and then the Kruskal–Wallis test was applied. This resampling operation is repeated 10,000 times. Table 3 reports the percentage of p-values less than 0.05 over the 10,000 runs.

Table 3. Percentage of p-value below the threshold of significance equal to 0.05 for the Kruskal–Wallis test of the pairwise comparison of Healthy and Extrapyramidal patients with the resampled Chronic patients.

Classes	Blinking Rate	Maximum Pupil Size Variation	Blinking Average Duration
H–C	59.11%	48.06%	10.99%
E–C	2.05%	67.92%	58.42%

The bootstrapping procedure allows us to validate the results in Table 1. Indeed, only the comparison between Healthy and Chronic patients with respect to the variation of pupil size has a percentage of significant p-values less than 50% (in particular equal to 48.06%), while this indicator has shown to be significant in the previous experiments. Therefore, we can conclude that, based on the considered indicators, healthy and extrapyramidal subjects look indistinguishable, while chronic pain patients behave significantly different. This is not an astonishing result as neurophysiological studies [37] suggested that a painful electrical stimulation is associated with consistent alterations in the eye muscle activity. Moreover, altered results of the Blink Reflex (BR) test normally stand for a dysfunction in brain stem and trigeminovascular connections of patients with migraine headache, supporting the trigeminovascular theory of migraine [38].

3.2. Mapping Latent Space Representations of ETT Images to Phenotypic Groups

For what concerns the analysis of the ETT dataset, three U–Net based autoencoders—one for each group, all sharing the same architecture and the same set of initial random weights—were trained for image reconstruction. In particular, the generic U–Net$_i$ is trained only on the data describing the ith class of individuals, which means that the U–Net$_H$ has been trained to reconstruct input images from the *healthy* class only, while U–Net$_E$ and U–Net$_C$ are trained on *extrapyramidal* and *chronic* classes, respectively. The workflow of the analysis is depicted in Figure 9.

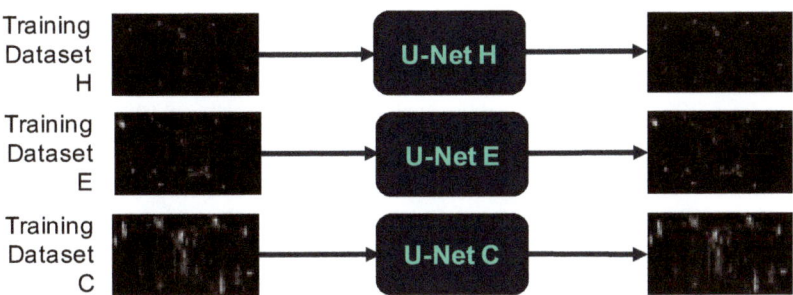

Figure 9. The three autoencoders trained for the three classes of subjects (chronic (C), healthy (H) and extrapyramidal (E)).

The experiments were carried out as described in the following. The three U-Net architectures were originally trained on 46 healthy, 46 extrapyramidal and 284 chronic pain patients, respectively, i.e., with the entire ETT dataset. Moreover, to obtain a balanced training set, experiments were also performed with only 46 chronic pain randomly sampled patients. The three encoder outputs were concatenated in a unique matrix, whose "entries" (latent representations of ETT images) were then clustered using the K-means algorithm, with $K = 3$.

The pipeline of the procedure is depicted in Figure 10. The number of clusters is empirically defined by the structure of the dataset itself, as it contains three types of individuals known a priori. The K-means algorithm is not used for classification purposes, but with the intent of evaluating the presence of useful information in the latent embeddings which allows to properly discern the three groups. In each of the three groups, only subjects belonging to the same cohort are present, showing the possibility of properly dividing patients into groups using the latent space embeddings.

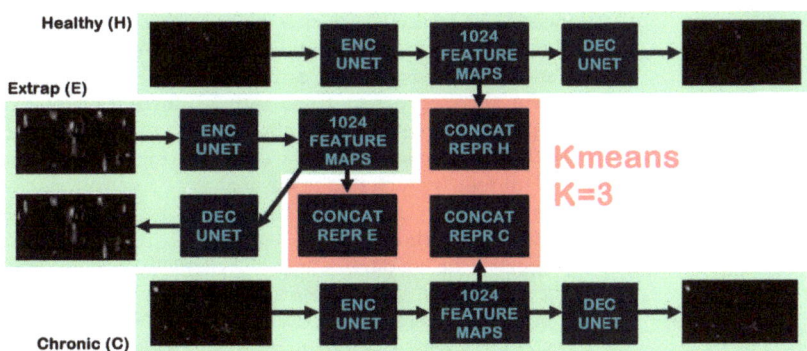

Figure 10. Graphical description of the pipeline for the extraction of latent image representations—via the U-Net—to which the K-means algorithm is applied.

Considering such preliminary results, we decided to implement a procedure to test the generalisation capability of the models. Therefore, we trained the three U-Nets only on 41 samples for each class of individuals. The test set, consisting of five healthy, five chronic and five extrapyramidal samples, was used as input to the three architectures separately. Subsequently, we ran the K-means algorithm ($K = 3$) with respect to the matrix obtained as the mean values along the embedding dimension of the test embeddings obtained at the previously described step. Next, we clustered the new mean values matrix, checking if the three clusters detected correspond to the three groups. In particular, the mean healthy embeddings obtained with the three architectures ended up in the same cluster, with 67% of accuracy. On the other hand, there was no remarkable distinction for the chronic and extrapyramidal patients. Moreover, as a further proof of concepts, we averaged the embedding representations, for each group of individuals and for each model—obtaining the vectors of the mean values for the reconstructed embeddings and clustering the corresponding matrix with K-means ($K = 3$) to verify whether the three averaged embeddings for the generic class could give an insight of the relationship between the input image class (healthy, chronic and extrapyramidal) in the embedding space. All of the three resulting mean "Healthy" reconstructed vectors of embeddings were clustered in the same community. Instead extrapyramidal and chronic patients, were not distinctively divided in their respective groups. This shows a similar behaviour to what we detected with the testing procedure. Healthy individuals trajectories are, in fact, more characterisable comparing to the two other subject categories.

Nonetheless, classifying the three groups of individuals based on DL techniques applied to ETT images remains a very difficult task, especially because of the scarcity of data and due to the complexity of the task itself. Indeed, human experts are unable to recognise different types of patients looking at the "frozen" trajectories they follow to approach the VSST, both because such trajectories are not so different to the naked eye and because, in the freezing process, the important temporal information on the way in which each trajectory is travelled, is irremediably lost. Taking into account, with an ad hoc preprocessing, of the sequential nature of the data and, most of all, enlarging the training dataset will surely allow better results.

4. Conclusions

In this paper, we presented some preliminary results on the analysis of VSST data, performed on three groups of individuals: patients affected by the extrapyramidal syndrome or by chronic pain symptoms and healthy subjects. Starting from the idea that the problem to be solved is multifaceted—which means that the data collected in a VSST have different nature and can be analysed from different viewpoints [22]—the goal of the present study is to detect if some regularities can be found within the data that allow to properly group them. Such detected differences could be potentially used in clinical practice, and therefore play an important role in evidencing possible neurological syndromes. The three-stage statistical analysis has been carried out on the basis of three metrics: the blinking rate, the maximum pupil size variation and the blinking average duration. The analysis showed the presence of some statistically significant differences between the groups analysed. In particular, the relevant difference in blinking rate between healthy and chronic patients is confirmed by each step of the analysis. Moreover, a statistical difference was detected between extrapyramidal and chronic patients for what concerns the maximum pupil size variation and blinking average duration. Conversely, based on the ETT (Eye-Tracking Trajectory) image dataset, a U-Net *ensemble* architecture was trained to reconstruct input images, using their latent representations, to appropriately cluster the visual data. Embeddings were, in fact, divided clearly into three separated groups. We performed preliminary testing, showing promising generalisation capabilities. Limitations of this work are mainly due to the small dataset available. Moreover variations of the VSST could be implemented and standardised, to avoid biases due to the fact that no instructions were given concerning the number of times the patients should have completed the sequence during the data acquisition time. Therefore, future research and extensions will concern new standardised data collection for further testing and a more extensive validation of the employed approaches based on a wider experimentation. For example a possible extension of the present study could be to consider more than three mutual exclusive classes, so as to include co-morbidities, i.e., cases in which additional conditions are concurrent to the primary one.

Author Contributions: Investigation, N.P., C.G., V.L. and G.M.D.; Conceptualisation and Methodology, N.P., C.G., V.L., M.B. and G.M.D.; Software, N.P., C.G., V.L. and G.M.D.; Supervision, M.L.S., M.B. and G.M.D.; Data Curation, E.Ș., N.P., C.G. and V.L.; Writing—original draft, N.P., C.G., V.L. and G.M.D.; and Writing—review and editing, N.P., C.G., V.L., M.L.S., M.B. and G.M.D. All authors have read and agreed to the published version of the manuscript.

Funding: This research received no external funding.

Institutional Review Board Statement: Not applicable.

Informed Consent Statement: The patients' consent was waived due the anonymous nature of the data.

Data Availability Statement: Not applicable.

Acknowledgments: The authors wish to thank RoNeuro Institute, part of the Romanian Foundation for the Study of Nanoneurosciences and Neuroregeneration, Cluj-Napoca, Romania, represented by Dafin Fior Muresanu, for providing the datasets used here for the experiments. Alessandra Rufa of the Department of Medicine, Surgery and Neuroscience, at the University of Siena, and Dario Zanca of the Department of Artificial Intelligence in Biomedical Engineering, at the University of Erlangen-Nürnberg, are also gratefully acknowledged for the fruitful discussions done at different stages of the present work.

Conflicts of Interest: The authors declare no conflict of interest.

References

1. Kredel, R.; Vater, C.; Klostermann, A.; Hossner, E.J. Eye-tracking technology and the dynamics of natural gaze behavior in sports: A systematic review of 40 years of research. *Front. Psychol.* **2017**, *8*, 1845. [CrossRef] [PubMed]
2. Verma, R.; Lalla, R.; Patil, T.B. Is blinking of the eyes affected in extrapyramidal disorders? An interesting observation in a patient with Wilson disease. In *Case Reports 2012*; BMJ Publishing Group: London, UK, 2012.
3. Medathati, N.V.K.; Ruta, D.; Hillis, J. Towards inferring cognitive state changes from pupil size variations in real world conditions. In Proceedings of the ACM Symposium on Eye Tracking Research and Applications, New York, NY, USA, 2–5 June 2020; Volume 22, pp. 1–10.
4. Zivkovic, M.; Bacanin, N.; Venkatachalam, K.; Nayyar, A.; Djordjevic, A.; Strumberger, I.; Al-Turjman, F. COVID-19 cases prediction by using hybrid machine learning and beetle antennae search approach. *Sustain. Cities Soc.* **2021**, *66*, 102669. [CrossRef]
5. Bacanin, N.; Stoean, R.; Zivkovic, M.; Petrovic, A.; Rashid, T.A.; Bezdan, T. Performance of a Novel Chaotic Firefly Algorithm with Enhanced Exploration for Tackling Global Optimization Problems: Application for Dropout Regularization. *Mathematics* **2021**, *9*, 2705. [CrossRef]
6. Malakar, S.; Ghosh, M.; Bhowmik, S.; Sarkar, R.; Nasipuri, M. A GA based hierarchical feature selection approach for handwritten word recognition. *Neural Comput. Appl.* **2020**, *32*, 2533–2552. [CrossRef]
7. Monaci, M.; Pancino, N.; Andreini, P.; Bonechi, S.; Bongini P.; Rossi, A.; Bianchini, M. Deep Learning Techniques for Dragonfly Action Recognition. In Proceedings of the ICPRAM, Valletta, Malta, 22–24 February 2020; pp. 562–569.
8. Mao, Y.; He, Y.; Liu, L.; Chen, X. Disease classification based on eye movement features with decision tree and random forest. *Front. Neurosci.* **2020**, *14*, 798. [CrossRef]
9. Vargas-Cuentas, N.I.; Roman-Gonzalez, A.; Gilman, R.H.; Barrientos, F.; Ting, J.; Hidalgo, D.; Zimic, M. Developing an eye-tracking algorithm as a potential tool for early diagnosis of autism spectrum disorder in children. *PLoS ONE* **2017**, *12*, e0188826. [CrossRef] [PubMed]
10. Duque, A.; Vázquez, C. Double attention bias for positive and negative emotional faces in clinical depression: Evidence from an eye-tracking study. *J. Behav. Ther. Exp. Psychiatry* **2015**, *46*, 107–114. [CrossRef] [PubMed]
11. Duque, A.; Vazquez, C. A failure to show the efficacy of a dot-probe attentional training in dysphoria: Evidence from an eye-tracking study. *J. Clin. Psychol.* **2018**, *74*, 2145–2160. [CrossRef] [PubMed]
12. Kellough, J.L.; Beevers, C.G.; Ellis, A.J.; Wells, T.T. Time course of selective attention in clinically depressed young adults: An eye tracking study. *Behav. Res. Ther.* **2008**, *46*, 1238–1243. [CrossRef] [PubMed]
13. Sanchez, A.; Vazquez, C.; Marker, C.; LeMoult, J.; Joormann, J. Attentional disengagement predicts stress recovery in depression: An eye-tracking study. *J. Abnorm. Psychol.* **2013**, *122*, 303. [CrossRef] [PubMed]
14. Hochstadt, J. Set-shifting and the on-line processing of relative clauses in Parkinson's disease: Results from a novel eye-tracking method. *Cortex* **2009**, *45*, 991–1011. [CrossRef] [PubMed]
15. Kaufmann, B.C.; Cazzoli, D.; Pflugshaupt, T.; Bohlhalter, S.; Vanbellingen, T.; Müri, R.M.; Nyffeler, T. Eyetracking during free visual exploration detects neglect more reliably than paper-pencil tests. *Cortex* **2020**, *129*, 223–235. [CrossRef]
16. Marx, S.; Respondek, G.; Stamelou, M.; Dowiasch, S.; Stoll, J.; Bremmer, F.; Einhauser, W. Validation of mobile eye-tracking as novel and efficient means for differentiating progressive supranuclear palsy from Parkinson's disease. *Front. Behav. Neurosci.* **2012**, *6*, 88. [CrossRef] [PubMed]

17. Trepagnier, C. Tracking gaze of patients with visuospatial neglect. *Top. Stroke Rehabil.* **2002**, *8*, 79–88. [CrossRef] [PubMed]
18. Davis, R. The Feasibility of Using Virtual Reality and Eye Tracking in Research With Older Adults With and Without Alzheimer's Disease. *Front. Aging Neurosci.* **2021**, *13*, 350. [CrossRef]
19. Maj, C.; Azevedo, T.; Giansanti, V.; Borisov, O.; Dimitri, G.M.; Spasov, S.; Merelli, I. Integration of machine learning methods to dissect genetically imputed transcriptomic profiles in alzheimer's disease. *Front. Genet.* **2019**, *10*, 726. [CrossRef] [PubMed]
20. Veneri, G.; Pretegiani, E.; Rosini, F.; Federighi, P.; Federico, A.; Rufa, A. Evaluating the human ongoing visual search performance by eye–tracking application and sequencing tests. *Comput. Methods Programs Biomed.* **2012**, *107*, 468–477. [CrossRef] [PubMed]
21. Veneri, G.; Pretegiani, E.; Fargnoli, F.; Rosini, F.; Vinciguerra, C.; Federighi, P.; Federico, A.; Rufa, A. Spatial ranking strategy and enhanced peripheral vision discrimination optimize performance and efficiency of visual sequential search. *Eur. J. Neurosci.* **2014**, *40*, 2833–2841. [CrossRef]
22. D'Inverno, G.A.; Brunetti, S.; Sampoli, M.L.; Mureşanu, D.F.; Rufa, A.; Bianchini, M. VSST analysis: An algorithmic approach. *Mathematics* **2021**, *9*, 2952. [CrossRef]
23. Ronneberger, O.; Fischer, P.; Brox, T. U-Net: Convolutional Networks for Biomedical Image Segmentation. *Lect. Notes Comput. Sci.* **2015**, *9351*, 234–241.
24. Bowie, H. Administration and interpretation of the Trail Making Test. *Nat. Protoc.* **2006**, *1*, 2277–2281. [CrossRef]
25. Carter, L. Best practices in eye tracking research. *Int. J. Psychophysiol.* **2020**, *155*, 49–62. [CrossRef] [PubMed]
26. Holmqvist, K.; Nyström, M.; Andersson, R.; Dewhurst, R.; Jarodzka, H.; Van de Weijer, J. *Eye Tracking: A Comprehensive Guide to Methods and Measures*; Oxford University Press: Oxford, UK, 2011.
27. The Mathworks, Inc. *MATLAB Version 9.10.0.1613233 (R2021a)*; The Mathworks, Inc.: Natick, MA, USA, 2021.
28. Kruskal, W.H.; Wallis, W.A. Use of ranks in one-criterion variance analysis. *J. Am. Stat. Assoc.* **1952**, *47*, 583–621. [CrossRef]
29. Vargha, A.; Delaney, H.D. The Kruskal-Wallis test and stochastic homogeneity. *J. Educ. Behav. Stat.* **1998**, *23*, 170–192. [CrossRef]
30. Amidan, B.G.; Ferryman, T.A.; Cooley, S.K. Data outlier detection using the Chebyshev theorem. In Proceedings of the 2005 IEEE Aerospace Conference, Big Sky, MT, USA, 5–12 March 2005; pp. 3814–3819.
31. Bianchini, M.; Dimitri, G.M.; Maggini, M.; Scarselli, F. Deep neural networks for structured data. In *Computational Intelligence for Pattern Recognition*; Springer: Cham, Switzerland, 2018; pp. 29–51.
32. LeCun, Y.; Bengio, Y.; Hinton, G. Deep learning. *Nature* **2015**, *521*, 436–444. [CrossRef] [PubMed]
33. Pancino, N.; Rossi, A.; Ciano, G.; Giacomini, G.; Bonechi, S.; Andreini, P.; Bongini, P. Graph Neural Networks for the Prediction of Protein-Protein Interfaces. In Proceedings of the ESANN, Bruges, Belgium, 2–4 October 2020; pp. 127–132.
34. Ronneberger, O.; Fischer, P.; Brox, T. U-net: Convolutional networks for biomedical image segmentation. In Proceedings of the International Conference on Medical Image Computing and Computer-Assisted Intervention, Munich, Germany, 5–9 October 2015; Springer: Cham, Switzerland, 2015.
35. Dimitri, G. M., Spasov, S., Duggento, A., Passamonti, L., & Toschi, N. (2020, July). Unsupervised stratification in neuroimaging through deep latent embeddings IEEE. In Proceedings of the 2020 42nd Annual International Conference of the IEEE Engineering in Medicine & Biology Society (EMBC) IEEE, Montreal, QC, Canada, 20–24 July 2020.
36. MacQueen, J. Some methods for classification and analysis of multivariate observations. In Proceedings of the Fifth Berkeley Symposium on Mathematical Statistics and Probability, Berkley, UK, 1 January 1967; Volume 1, pp. 281–297.
37. Peddireddy, A.; Wang, K.; Svensson, P.; Arendt-Nielsen, L. Blink reflexes in chronic tension–type headache patients and healthy controls. *Clin. Neurophysiol.* **2009**, *120*, 1711–1716. [CrossRef] [PubMed]
38. Unal, Z.; Domac, F.M.; Boylu, E.; Kocer, A.; Tanridag, T.; Us, O. Blink reflex in migraine headache. *North. Clin. Istanb.* **2016**, *3*, 289–292.

Article

A Multi-Stage GAN for Multi-Organ Chest X-ray Image Generation and Segmentation

Giorgio Ciano [1,2,*], Paolo Andreini [2], Tommaso Mazzierli [3], Monica Bianchini [2] and Franco Scarselli [2]

1. Department of Information Engineering, University of Florence, 50121 Florence, Italy
2. Department of Information Engineering and Mathematics, University of Siena, 53100 Siena, Italy; paolo.andreini@unisi.it (P.A.); monica.bianchini@unisi.it (M.B.); franco@diism.unisi.it (F.S.)
3. Department of Nephrology, AOU Careggi, University of Florence, 50121 Florence, Italy; tommaso.mazzierli@unifi.it
* Correspondence: giorgio.ciano@unifi.it

Citation: Ciano, G.; Andreini, P.; Mazzierli, T.; Bianchini, M.; Scarselli, F. A Multi-Stage GAN for Multi-Organ Chest X-ray Image Generation and Segmentation. *Mathematics* **2021**, *9*, 2896. https://doi.org/10.3390/math9222896

Academic Editor: Ezequiel López-Rubio

Received: 9 October 2021
Accepted: 11 November 2021
Published: 14 November 2021

Publisher's Note: MDPI stays neutral with regard to jurisdictional claims in published maps and institutional affiliations.

Copyright: © 2021 by the authors. Licensee MDPI, Basel, Switzerland. This article is an open access article distributed under the terms and conditions of the Creative Commons Attribution (CC BY) license (https://creativecommons.org/licenses/by/4.0/).

Abstract: Multi-organ segmentation of X-ray images is of fundamental importance for computer aided diagnosis systems. However, the most advanced semantic segmentation methods rely on deep learning and require a huge amount of labeled images, which are rarely available due to both the high cost of human resources and the time required for labeling. In this paper, we present a novel multi-stage generation algorithm based on Generative Adversarial Networks (GANs) that can produce synthetic images along with their semantic labels and can be used for data augmentation. The main feature of the method is that, unlike other approaches, generation occurs in several stages, which simplifies the procedure and allows it to be used on very small datasets. The method was evaluated on the segmentation of chest radiographic images, showing promising results. The multi-stage approach achieves state-of-the-art and, when very few images are used to train the GANs, outperforms the corresponding single-stage approach.

Keywords: deep learning; convolutional neural networks; semantic segmentation; generative adversarial networks; chest X-ray; image augmentation

1. Introduction

Chest X-ray (CXR) is one of the most used techniques worldwide for the diagnosis of various diseases, such as pneumonia, tuberculosis, infiltration, heart failure and lung cancer. Chest X-rays have enormous advantages: they are cheap, X-ray equipment is also available in the poorest areas of the world and, moreover, the interpretation/reporting of X-rays is less operator-dependent than the results of other more advanced techniques, such as computed tomography (CT) and magnetic resonance (RMI). Furthermore, undergoing this examination is very fast and minimally invasive [1]. Recently, CXR images have gained even greater importance due to COVID-19, which mainly causes lung infection and, after healing, often leaves widespread signs of pulmonary fibrosis: the respiratory tissue affected by the infection loses its characteristics and its normal structure. Consequently, CXR images are often used for the diagnosis of COVID-19 and for treatment of the aftereffects of SARS-CoV-2 [2–4].

Therefore, with the rapid growth in the number of CXRs performed per patient, there is an ever-increasing need for computer-aided diagnosis (CAD) systems to assist radiologists, since manual classification and annotation is time-consuming and subject to errors. Recently, deep learning (DL) has radically changed the perspective in medical image processing, and deep neural networks (DNNs) have been applied to a variety of tasks, including organ segmentation, object and lesion classification [5], image generation and registration [6]. These DL methods constitute an important step towards the construction of CADs for medical images and, in particular, for CXRs.

Semantic segmentation of anatomical structures is the process of classifying each pixel of an image according to the structure to which it belongs. In CAD, segmentation plays a fundamental role. Indeed, segmentation of CXR images is usually necessary to obtain regions of interest and allows the extraction of size measurements of organs (e.g., cardiothoracic ratio quantification) and irregular shapes, which can provide meaningful information on important diseases, such as cardiomegaly, emphysema and lung nodules [7]. Segmentation may also help to improve the performance of automatic classification: in [8], it is shown that, by exploiting segmentation, DL models focus their attention primarily on the lung, not taking into account unnecessary background information and noise.

Modern state-of-the-art segmentation algorithms are largely based on DNNs [9–11]. However, to achieve good results, DNNs need a fairly large amount of labeled data. Therefore, the main problem with segmentation by DNNs is the scarce availability of appropriate datasets to help solve a given task. This problem is even more evident in the medical field, where data availability is affected by privacy concerns and where a great deal of time and human resources are required to manually label each pixel of each image.

A common solution to cope with this problem is the generation of synthetic images, along with their semantic label maps. This task can be carried out by Generative Adversarial Networks (GANs) [12], which can learn, using few training examples, the data distribution in a given domain. In this paper, we present a new model, based on GANs, to generate multi-organ segmentation of CXR images. Unlike other approaches, the main feature of the proposed method is that generation occurs in three stages. In the first stage, the position of each anatomical part is generated and represented by a "dot" within the image; in the second stage, semantic labels are obtained from the dots; finally, the chest X-ray image is generated. Each step is implemented by a GAN. More precisely, we adopt Progressively Growing GANs (PGGANs) [13], a recent extension of GANs that allows the generation of high resolution images, and Pix2PixHD [14] for the translation steps. The intuitive idea underlying the approach is that generation benefits by the multi-stage procedure, since the GAN used in each single step faces a subproblem, and can be trained using fewer data. Actually, the generalization capability of neural networks, and more generally of deep learning approaches, has a solid mathematical foundation (see, e.g., the seminal work [15] and the more recent papers [16,17]). The most general rule states that the simpler the model the better its generalization capability. In our approach, the simplification lies in that, in the three-stage method, the tasks to be solved in each of the three steps are simpler and require less effort.

In order to evaluate the performance of the proposed method, synthetic images were used to train a segmentation network (here, we use the Segmentation Multiscale Attention Network (SMANet) [18], a deep convolutional neural network based on the Pyramid Scene Parsing Network [11]), subsequently applied to a popular benchmark for multi-organ chest segmentation, the Segmentation in Chest Radiographs (SCR) dataset [6]. The results obtained are very promising and exceed (to the best of our knowledge) those obtained by other previous methods. Moreover, the quality of the produced segmentation was confirmed by physicians. Finally, to demonstrate the capabilities of our approach, especially having little data available, we compared it to two other methods, using only 10% of the images in the dataset. In particular, the multi-stage approach was compared with a single-stage method—in which chest X-ray images and semantic label maps are generated simultaneously—and with a two-stage method—where semantic label maps are generated and then translated into X-ray images. The experimental results show that the proposed three-stage method outperforms the two-stage method, while the two-stage overcomes the single-stage approach, confirming that splitting the generation procedure can be advantageous, particularly when few training images are available.

The paper is organized as follows. In Section 2, the related literature is reviewed. Section 3 presents a description of the proposed image generation method. Section 4 shows and discusses the experimental results. Finally, in Section 5, we draw some conclusions and describe future research.

2. Related Works

In the following, recent works related to the topics addressed in this paper are briefly reviewed, namely regarding synthetic image generation, image-to-image translation, and the segmentation of medical images.

2.1. Synthetic Image Generation

Methods for generating images are by no means new and can be classified into two main categories: model-based and learning-based approaches. A model-based method consists of formulating a model of the observed data to render the image by a dedicated engine. This approach has been widely adopted to generate images in many different domains [19–21]. Nonetheless, the design of specialized engines for data generation requires a deep knowledge of the specific domain. For this reason, in recent years, the learning-based approach has attracted increasing research interest. In this context, machine learning techniques are used to capture the intrinsic variability of a set of training images, so that the specific domain model is acquired implicitly from the data. Once the probability distribution that underlies the set of real images has been learned, the system can be used to generate new images that are likely to mimic the original ones. One of the most successful machine learning models for data generation is the Generative Adversarial Network (GAN) [12]. A GAN is composed by two networks: a generator G and a discriminator D. The former learns to generate data starting from a latent random variable $\mathbf{z} \in \mathbb{R}^Z$, while the latter aims at distinguishing real data from generated ones. Training GANs is difficult, because it consists of a min-max game between two neural networks and convergence is not guaranteed. This problem is compounded in the generation of high resolution images, because the high resolution makes it easier to distinguish generated images from training images [22]. One of the most successful approaches to face this problem is represented by Progressively Growing GANs (PGGANs) [13]. This model, in fact, is based on a multi-stage approach that aims to simplify and stabilize the training and allows it to generate high resolution images. More specifically, in a PGGAN, the training starts at low resolution, while new blocks are progressively introduced into the system to increase the resolution of the generation. The generator and discriminator grow symmetrically until the desired resolution is reached. Based upon PGGANs, many different approaches have been proposed. For instance, StyleGANs [23] maintain the same discriminator as PGGANs, but introduce a new generator which is able to control the style of the generated images at different levels of detail. In StyleGAN2s [24], an improved training scheme is introduced, which achieves the same goal—training starts by focusing on low resolution images and then progressively shifts the focus to higher and higher resolutions—without changing the network topology during training. In this way, the updated model shows improved results at the expense of longer training times and more computing resources.

In this paper, we use PGGANs in three different ways. For the single-stage method, a PGGAN simultaneously generates semantic label maps and CXR images. For the two-stage method, only semantic label maps are generated, while for the three-stage method we use a PGGAN to generate "dots" that correspond to different anatomical parts.

2.2. Image-to-Image Translation

Recently, besides image generation, adversarial learning has also been employed for image-to-image translation, the goal of which is to translate an input image from one domain to another. Many computer vision tasks, such as image super-resolution [25], image inpainting [26], and style transfer [27], can be cast into the image-to-image translation framework. Both unsupervised [28–31] and supervised approaches [13,32,33] can be used but, for the proposed application to CXR image generation, the unsupervised category is not relevant. Supervised training uses a set of pairs of corresponding images $\{(s_i, t_i)\}$, where s_i is an image of the source domain and t_i is the corresponding image in the target domain. In the original GAN framework, there is no explicit way of controlling what to generate, since the output depends only on the latent vector $\mathbf{z}$. For this reason, in conditional GANs (cGANs) [34], an additional input **c** is introduced to guide the generation. In a cGAN, the generator can be defined accordingly as $G(\mathbf{c}, \mathbf{z})$. Pix2Pix [32] is a general approach for image-to-image translation and consists of a conditional GAN that operates in a supervised way. Pix2Pix uses a loss function that allows it to generate plausible images in relation to the destination domain, which are also credible translations of the input image. With respect to supervised image-to-image translation techniques, in addition to the aforementioned Pix2Pix, the most used models are CRN [33], Pix2PixHD [14], BycicleGAN [35], SIMS [36], and SPADE [37]. In particular, Pix2PixHD [14] improves upon Pix2Pix by employing a coarse-to-fine generator and discriminator, along with a feature-matching loss function, allowing it to translate images with higher resolution and quality.

For the image-to-image translation phase, we use the Pix2PixHD network. The single-stage method does not require a translation step, while for the two-stage method we use Pix2PixHD to obtain a CXR image from the label map. Finally, in the three-stage method, Pix2PixHD is used in two steps: for the translation from "dots" to semantic label maps and, after that, for the translation of label maps into CXR images.

2.3. Medical Image Generation

In recent years, GANs have attracted the attention of medical researchers, their applications ranging from object detection [38–40] to registration [41–43], classification [44–46] and segmentation [47,48] of images. For instance, in [49], different GANs have been used for the synthesis of each class of liver lesion (cysts, metastases and hemangiomas). However, in the medical domain, the use of complex machine learning models is often limited by the difficulty of collecting large sets of data. In this context, GANs can be employed to generate synthetic data, realizing a form of data augmentation. In fact, GAN generated data can be used to enlarge the available datasets and improve the performance in different tasks. As an example, GAN generated images have been successfully used to improve the performance in classification problems, by combining real and synthetic images during the training of a classifier. In [50], Wasserstein GANs (WGANs) and InfoGANs have been combined to classify histopathological images, whereas in [44] WGAN and CatGAN generated images were used to improve the classification of dermoscopic images. Only in a few cases have GANs been used to generate chest radiographic images, as in [45], where images for cardiac abnormality classification were obtained with a semi-supervised architecture, or in [51], where GANs were used to generate low resolution (64×64) CXRs to diagnose pneumonia. More related to this work, in [19], high-resolution synthetic images of the retina and the corresponding semantic label maps have been generated. Moreover, synthesizing images has been proven to be an effective method for data augmentation, that can be used to improve performance in retinal vessel segmentation.

In this paper, chest X-ray images were generated with the corresponding semantic label maps (which correspond to different organs). We then used such images to train a segmentation network, with very promising results.

2.4. Organ Segmentation

X-rays are one of the most used techniques in medical diagnostics. The reasons are medical and economic, since they are cheap, noninvasive and fast examinations. Many diseases, such as pneumonia, tuberculosis, lung cancer, and heart failure are commonly diagnosed from CXR images. However, due to overlapping organs, low resolution and subtle anatomical shape and size variations, interpreting CXRs accurately remains challenging and requires highly qualified and trained personnel. Therefore, it is of a great clinical and scientific interest to develop computer-based systems that support the analysis of CXRs. In [52], a lung boundary detection system was proposed, building an anatomical atlas to be used in combination with graph cut-based image region refinement [53–55]. A method for lung field segmentation, based on joint shape and appearance sparse learning, was proposed in [56], while a technique for landmark detection was presented in [57]. Haar-like features and a random forest classifier were combined for the appearance of landmarks. Furthermore, a Gaussian distribution augmented by shape-based random forest classifiers was adopted for learning spatial relationships between landmarks. *InvertedNet*, an architecture able to segment the heart, clavicles and lungs, was introduced in [58]. This network employs a loss function based on the Dice Coefficient, Exponential Linear Units (ELUs) activation functions, and a model architecture that aims at containing the number of parameters. Moreover, the UNet [59] architecture has been widely used for lung segmentation, as in [60–62]. In the Structure Correcting Adversarial Network (SCAN) [63] a segmentation network and a critic network were jointly trained with an adversarial mechanism for organ segmentation in chest X-rays.

3. Chest X-ray Generation

The main goal of this study is to prove that by dividing the generation problem into multiple simpler stages, the quality of the generated images improves, so that they can be more effectively employed as a form of data augmentation. More specifically, we compare three different generation approaches. The first method, described in Section 3.1, consists of generating chest X-ray images and the corresponding label maps in a single stage. In the second approach, presented in Section 3.2, the generation procedure is divided into two stages, where the label maps are initially generated and then translated into images. The third method, reported in Section 3.3, consists of a three-stage approach, that starts by generating the position of the objects in the image, then the label maps and, finally, the X-ray images. The images generated employing each of the three approaches are comparatively evaluated by training a segmentation network.

To increase the descriptive power of real images, especially with regards to the position of the various organs, standard data augmentation has preventively been applied. Therefore, the original X-ray images, along with their corresponding masks, were augmented by applying random rotations in the interval $[-2, 2]$ degrees, random horizontal, vertical and combined translations from -3% to $+3\%$ of the number of pixels, and adding a Gaussian noise—only to the original images—with a zero mean and variance between 0.01 and 0.03×255. For the generation of images, we essentially used two networks well known in the literature, namely PGGANs [13] and Pix2PixHD [14], and their details are given in the following sections. In particular, in Sections 3.1–3.3, we extensively describe the three different generation procedures, respectively the single-stage, two-stage and three-stage methods. The next Section 3.4 presents the semantic segmentation network that was employed. Finally, some details on the training method are collected in Section 3.5.

3.1. Single-Stage Method

This baseline approach consists of stacking X-ray images and labels into two different channels, which are simultaneously fed into the PGGAN. Therefore, the PGGAN is trained to generate pairs composed by an X-ray image and its corresponding label (see Figure 1).

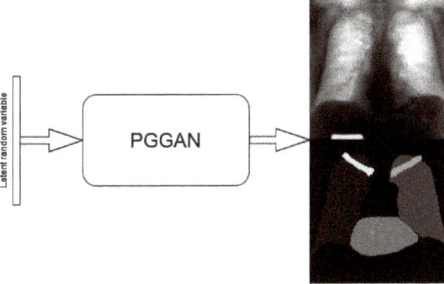

Figure 1. The one-stage image generation scheme. The input of the network is a latent vector, while the PGGAN simultaneously produces the label map and the X-ray image.

3.2. Two-Stage Method

In this approach, the generation procedure is divided into two steps. The first one consists of generating the labels through a PGGAN, while, in the second, the translation from the label to the corresponding chest X-ray image is carried out using Pix2PixHD (see Figure 2).

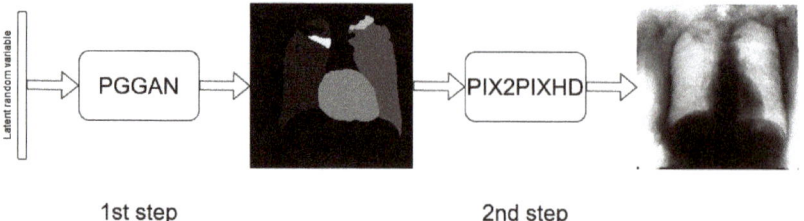

Figure 2. The two-stage image generation scheme. In the first step, the PGGAN takes in input as a latent vector and produces the label map. The generated label map is then used as input to a Pix2PixHD module, which is trained to output the X-ray image.

3.3. Three-Stage Method

It consists of further subdividing the generation procedure, with a first phase consisting of generating the position and type of the objects that will be generated later, regardless of their shape or appearance. This is obtained by generating label maps that contain "dots" in correspondence with different anatomical parts (lungs, heart, clavicles). The dots can be considered as "seeds", from which, through the subsequent steps, the complete label maps are realized (second phase). Finally, in the last step, chest X-ray images are generated from the label maps. The exact procedure is described in the following. Initially, label maps containing "dots", with a specific value for each anatomical part, are created. The position of the "dot" center is given by the centroid of each labeled anatomical part. The label maps generated in this phase have a low resolution (64 × 64), as a high level of detail is not necessary, because the exact object shapes are not defined—but only their centroid positions. It should be observed that this also allows a significant reduction in the computational burden of this stage and speeds up the computation. The generated label maps must be subsequently resized to the original image resolution—required in the following stages of generation (a nearest neighbour interpolation was used to maintain the original label codes)—and translated into labels, which will be finally translated into images, using Pix2PixHD (see Figure 3).

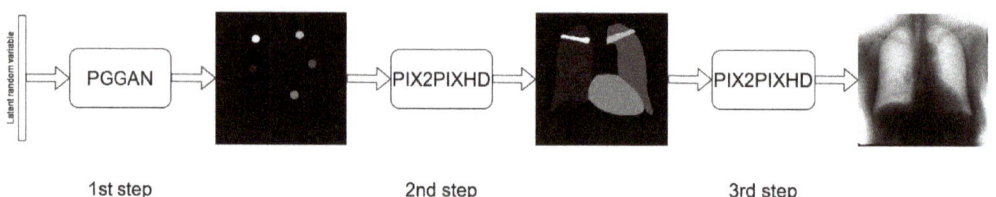

Figure 3. The three-stage image generation scheme. In the first step, dots are generated from a latent vector. Then, Pix2PixHD translates dots into a label map, and finally the label map is translated into an X-ray image.

3.4. Segmentation Multiscale Attention Network

After generating the label maps of the corresponding chest X-ray images, we use a semantic segmentation network to prove the effectiveness of the synthetic images during training, and to compare the three-stage approach with the one- and two-stage methods, proving its superior performance. In this paper, the Segmentation Multiscale Attention Network (SMANet) [18] was employed. The SMANet is composed of three main components, a ResNet encoder, a multi-scale attention module, and a convolutional decoder (see Figure 4).

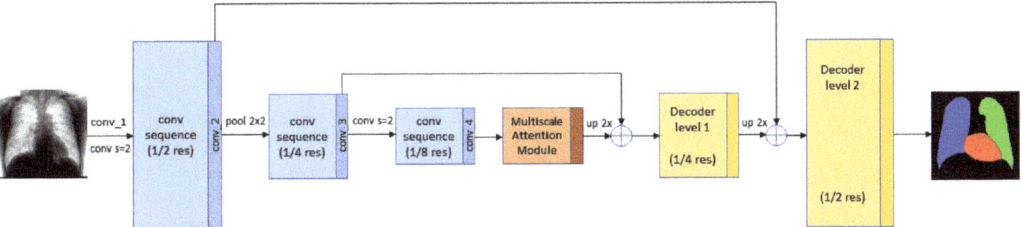

Figure 4. Scheme of the SMANet segmentation network.

This architecture, initially proposed for scene text segmentation, is based on the Pyramid Scene Parsing Network (PSPNet) [11], a deep fully convolutional neural network with a ResNet [64] encoder. Dilated convolutions (i.e. atrous convolutions [65]) are used in the ResNet backbone, to widen the receptive field of the neural network in order to avoid an excessive reduction of the spatial resolution due to down-sampling. The most characteristic part of the PSPNet architecture is the pyramid pooling module (PSP), which is employed to capture features of different scale in the image. In the SMANet, the PSP module is replaced with a multi-scale attention mechanism to better focus on the relevant objects present in the image. Finally, a two-level convolutional decoder is added to the architecture to improve the recognition of small objects.

3.5. Training Details

The PGGAN architecture, proposed in [13], was employed for image generation; the number of parameters were modified to speed up learning and reduce overfitting. More specifically, the maximum number of feature maps for each layer was reduced to 64. Furthermore, since the PGGAN was used to generate seeds and labels, obtaining only the semantic label maps in both cases, the output image has only one channel instead of three. The generation procedure (PGGAN and Pix2PixHD) was stopped by visually examining the generated samples during the training phase. The images, generated in the various steps for all the methods, have a resolution of 1024×1024, except in the case of the "dot" label maps, which, as mentioned before, are generated at a 64×64 resolution.

The SMANet is implemented in TensorFlow. Random crops of 377×377 pixels were employed during training, whereas a sliding window of the same size was used for testing. The Adam optimizer [66], based on an initial learning rate of 10^{-4} and a mini batch of 17 examples, was used to train the SMANet. All the experiments were carried out in a Linux environment on a single NVIDIA Tesla V100 SXM2 with 32 GB RAM. The SMANet's goal is to produce the semantic segmentation of the lungs and heart. The network is trained by a supervised approach, both in the case of real and synthetic images. In particular, for the images generated by the three different methods, we are able to use this approach thanks to the generation of both the images and the label maps.

4. Experiments and Results

In this section, after describing the dataset on which our new proposed method was tested, we evaluate the results obtained, both qualitatively—based on the judgment of three physicians—and quantitatively, comparing them with related approaches present in the literature.

4.1. Dataset

Chest X-ray images are available thanks to the Japanese Society of Radiological Technology (JSRT) [67]. The dataset they provide consists of 247 chest X-ray images. The resolution of the images is 2048 × 2048 pixels, with a spatial resolution of 0.175 mm/pixel and 12 bit gray levels. Furthermore, segmentation supervisions for the JSRT database are available in the Segmentation in the Chest Radiographs (SCR) dataset [6]. More precisely, this dataset provides chest X-ray supervisions which correspond with the pixel-level positions of the different anatomical parts. Such supervisions were produced by two observers who segmented five objects in each image: the two lungs, the heart and the two clavicles. The first observer was a medical student and his segmentation was used as the gold standard, while the second observer was a computer science student, specialized in medical imaging, and his segmentation was considered that of a human expert.

The SCR dataset comes with an official splitting, which is employed in this paper and consists of 124 images for learning and 123 for testing. We use two different experimental configurations. In the former, called FULL_DATASET, all the training images are exploited. More precisely, the PGGAN generation network is trained on the basis of 744 images, available in the SCR training set and obtained with the augmentation procedure described above. The SMANet is trained on 7500 synthetic images, generated by the PGGAN, and fine-tuned on the 744 images extracted from the SCR training set, while 2500 synthetic images are used for validation. For the second configuration, called TINY_DATASET, only 10% of the SCR training set is used and the PGGAN is trained on only 66 images (obtained both from SCR and with augmentation); furthermore, the SMANet is trained exactly as above, except for the fine-tuning, which is carried out on 66 images.

4.2. Quantitative Results

Generated images were employed to train a deep semantic segmentation network. The rationale behind the approach is that the performance of the network trained on the generated data reflects the data quality and variety. A good performance of the segmentation network indicates that the generated data successfully capture the true distribution of the real samples. To assess the segmentation results, some standard evaluation metrics were used. The Jaccard Index, J, also called Intersection Over Union (IOU), measures the similarity between two finite sample sets—the predicted segmentation and the target mask in this case—and is defined as the size of their intersection divided by the size of their union. For binary classification, the Jaccard index can be framed in the following formula:

$$J = \frac{TP}{TP + FP + FN}$$

where TP, FP and FN denote the number of true positives, false positives and false negatives, respectively. Furthermore, the Dice Score, DSC, is defined as:

$$DSC = \frac{2 \times TP}{2 \times TP + FP + FN}$$

DSC is a *quotient of similarity between sets* and ranges between 0 and 1.

The experiments can be divided into two phases: first, we evaluated the generation procedure described in Section 3.3 using the FULL_DATASET, then, we compared this approach with the other two methods described in Sections 3.1 and 3.2 using the TINY_DATASET. The purpose of this latter experiment was to evaluate whether multi-stage generation methods are actually more effective in producing data suitable for semantic segmentation with a limited amount of data. In particular, in the experimental setup based on the FULL_DATASET, for the three-stage method, the generation network was trained on all the SCR training images, to which the augmentation procedure described in Section 3 was applied. Then, 10,000 synthetic images were generated and used to train the semantic segmentation network. Moreover, we evaluated a fine-tuning of the network on the SCR real images after the pre-training on the generated images. The results, shown in Table 1, are compared with those obtained using only real images to train the semantic segmentation network, which can be considered as a baseline.

Table 1. Evaluation of the proposed methods based on the FULL_DATASET, using 2500 generated images for the validation set. **Real** corresponds to the results obtained using the official training set; *Synth 3* corresponds to the results obtained using only the generated images, while in the *Finetune* column, real data are employed for fine-tuning.

		Real	Three-Stage	
			Synth 3	Finetune
J	Left Lung	96.10	95.30	**96.22**
	Heart	90.78	87.25	**91.11**
	Right Lung	**96.85**	96.15	96.79
	Average	94.58	92.90	**94.71**
DSC	Left Lung	98.01	97.6	**98.07**
	Heart	95.17	93.19	**95.35**
	Right Lung	**98.40**	98.04	98.37
	Average	97.19	96.28	**97.26**

Next, the TINY_DATASET was used in order to evaluate the performance of the methods with a very small dataset. More precisely, the following experimental setups, the results of which are shown in Table 2, are considered:

- REAL—only real images are used for training the semantic segmentation network;
- SINGLE-STAGE—the segmentation network uses the images generated by the single-stage method (Synth 1 in the tables) for training while real images are employed for fine-tuning (Finetune in the tables);
- TWO-STAGES—the images generated with the two-stage method are used to pre-train the segmentation network (Synth 2) while real images are used for fine-tuning;
- THREE-STAGE—the images generated with the three-stage method are used for training the segmentation network (Synth 3), while real images are employed for fine-tuning.

In this case, the PGGAN was trained on 66 images, based on 11 images randomly chosen from the entire training set to which the augmentation described above was applied.

Table 2. Evaluation of the proposed methods based on the TINY_DATASET, using 2500 generated images for the validation set. **Real** corresponds to the results obtained using the official training set; *Synth 1, Synth 2, Synth 3*, correspond to the results obtained using only the generated images, while in the *Finetune* columns, real data are employed for fine-tuning.

		Real	Single-Stage		Two-Stage		Three-Stage	
			Synth 1	Finetune	Synth 2	Finetune	Synth 3	Finetune
J	Left Lung	93.70	55.59	74.11	94.91	94.4	94.96	**95.29**
	Heart	85.50	0.07	37.47	86.98	85.21	87.27	**87.47**
	Right Lung	93.70	52.78	79.99	95.90	95.44	95.90	**95.92**
	Average	90.97	36.15	63.86	92.60	91.68	92.71	**92.89**
DSC	Left Lung	96.75	71.46	85.13	97.39	97.12	97.42	**97.59**
	Heart	92.18	0.13	54.51	93.04	92.02	93.20	**93.32**
	Right Lung	96.74	69.09	88.89	97.91	97.66	97.90	**97.92**
	Average	95.22	46.89	76.18	96.11	95.60	96.17	**96.28**

In general, we can see that the best results are obtained with the three-stage method followed by fine-tuning. From Table 1, we observe a small improvement in results using a fine-tune on a network previously trained with images generated using the three-stage method. Therefore, the three-stage method provides good synthetic data, but the advantage given by generated images is low when the training set is large. Conversely, when few training images are available, in the TINY_DATASET setup, multi-stage methods outperform the baseline (column REAL of Table 2) and this happens even without fine-tuning. Thus, in this case, the advantage provided by synthetic images is evident. Moreover, the three-stage method outperforms the two-stage approach, even with fine-tuning, which confirms our claim that splitting the generation procedure may provide a performance increase when few training images are available.

Finally, it is worth noting that fine-tuning improves the performance of the three-stage method, both in the FULL_DATASET and in the TINY_DATASET framework, which does not hold for the two-stage method. This behaviour may be explained by some complementary information that is captured from real images only with the three-stage method. Actually, we may argue that, in different phases of a multi-stage approach, different types of information can be captured: such a diversification seems to provide an advantage to the three-stage method, which develops some capability to model the data domain with more orthogonal information.

4.3. Comparison with Other Approaches

Table 3 shows our best results and the segmentation performance published by all recent methods, of which we are aware, on the SCR dataset. According to the results in the table, the three-stage method obtained the best performance score both for the lungs and the heart.

However, it is worth mentioning that Table 3 gives only a rough idea of the state-of-the-art, since a direct comparison between the proposed method and other approaches is not feasible, our primary focus being on image generation, in contrast with the comparative approaches that are mainly devoted to segmentation, and for which no results are reported on small image datasets. Moreover, the previous methods used different partitions of the SCR dataset to obtain the training and the test set, such as two-fold, three-fold, five-fold cross-validation or ad hoc splittings, which are often not publicly available, while, in our experiments, we preferred to use the original partition, provided with the SCR dataset (note that, compared to most of the other solutions used in comparative methods, the original subdivision has the disadvantage of producing a smaller training set, which is not in conflict, however, with the purpose of the present work). Finally, a variety of different image sizes have also been used, ranging from 256×256, to 400×400, and to 512×512—the resolution used in this work.

Table 3. Comparison of segmentation results among different methods on the SCR dataset (CV stands for cross-validation).

Method	Image Size	Augmentation	Evaluation Scheme	Lungs		Heart	
				DSC	J	DSC	J
Human expert [6]	2048×2048	No	-	-	94.6	-	87.8
U-Net [60]	256×256	No	5-fold CV	-	95.9	-	89.9
InvertedNet [58]	256×256	No	3-fold CV	97.4	95	93.7	88.2
SegNet [62]	256×256	No	5-fold CV	97.9	95.5	94.4	89.6
FCN [62]	256×256	No	5-fold CV	97.4	95	94.2	89.2
SCAN [58]	400×400	No	training/test split (209/38)	97.3	94.7	92.7	86.6
Our three-stage method	512×512	Yes	official split	**98.2**	**96.5**	**95.36**	**91.1**

4.4. Qualitative Results

In this section, some examples of images and corresponding segmentations, generated with the approaches described in Section 3, are qualitatively examined. We also report some comments from three physicians on the generated segmentations, to provide a medical assessment of the quality of our method.

Figures 5 and 6 display some examples—randomly chosen from all the generated images—of the label maps and the corresponding chest X-ray images generated with the three methods described in Section 3, using the FULL_DATASET and the TINY_DATASET, respectively. We can observe that, with the single and two-stage methods, the images tend to be more similar to those belonging to the training set. For example, in most of the generated images there are white rectangles, which resemble those present in the training images, used to cover the names of both the patient and the hospital. Instead, the three-stage method does not produce such artifacts, suggesting that it is less prone to overfitting.

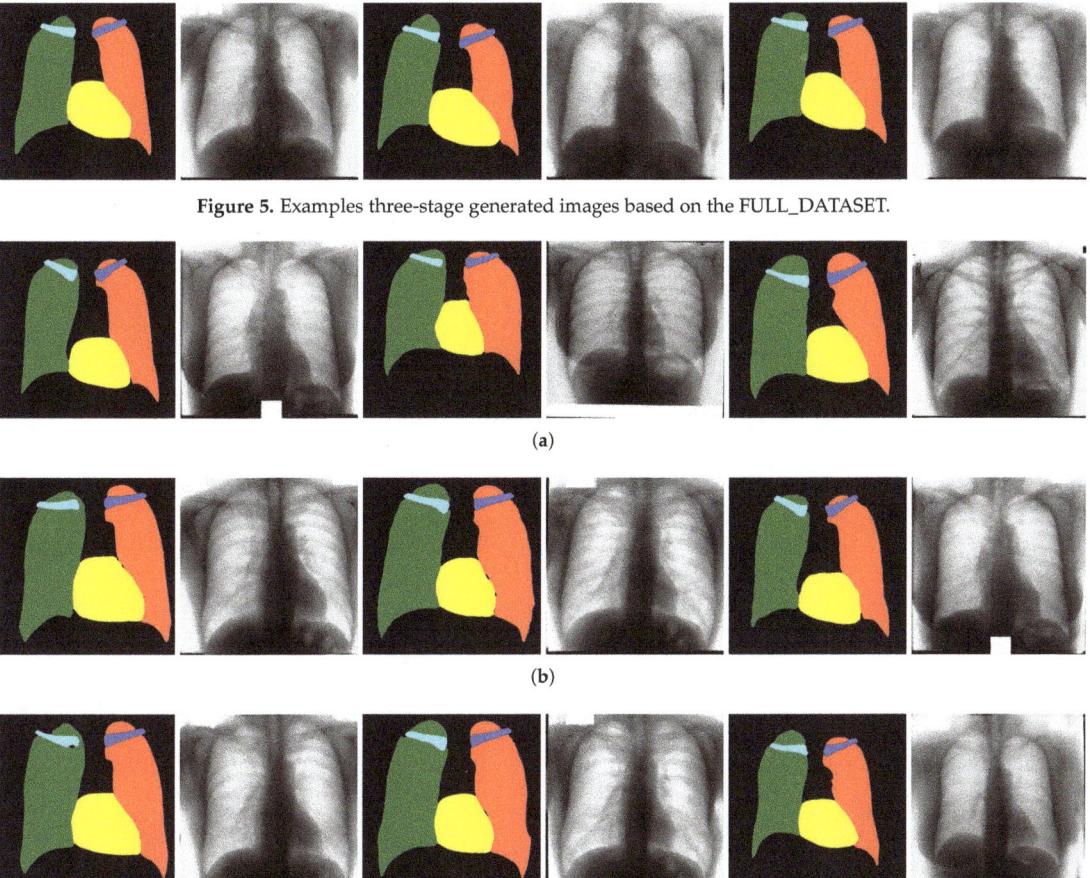

Figure 5. Examples three-stage generated images based on the FULL_DATASET.

Figure 6. Examples of generated images based on the TINY_DATASET. (**a**) Single-stage 10% of generated images, (**b**) Two-stage 10% of generated images, (**c**) Three-stage 10% of generated images.

Moreover, in order to clarify the limits of the three-stage method, we assessed the quality of the segmentation results based on three human experts, who were asked to check 20 chest X-ray images, along with the corresponding supervision and the segmentation obtained by the SMANet network. Such images were chosen among those that can be considered difficult, at least based on the high error obtained by the segmentation algorithm. Figures 7 and 8 show different examples of the images evaluated by the experts. The first column represents the chest X-ray image, while the second and the third columns, the order of which was randomly exchanged during the presentation to the experts, represent the target segmentation and our prediction, respectively. The three physicians were asked to choose the best segmentation and to comment about their choice. Apart from a general agreement of all the doctors on the good quality of both the target segmentation and the segmentation provided by the three-stage method, surprisingly, they often chose the second one. For the examples in Figure 7, for instance, all the experts shared the same opinion, preferring the segmentation obtained by the SMANet over the ground-truth segmentation. To report the results of the qualitative analysis, we numbered the target and predicted segmentation with numbers 1 and 2, respectively, while doctors were assigned unordered pairs to obtain an unbiased result. Then, with respect to Figure 7a, the comments reported by the experts were: (1) In segmentation 1, a fairly large part of the upper left ventricle is missing; (2) I choose the segmentation number 2 because the heart profile does not protrude to the left of the spine profile; (3) The best is number 2, the other leaves out a piece of the left free edge of the heart, in the cranial area. Furthermore, for Figure 7b, we obtained: (1) The second image is the best for the cardiac profile. For lung profiles, the second image is always better. The only flaw is that it leaks a bit on the right and left costophrenic sinuses. (2) Image 2 is the best, because the lower cardiac margin is lying down and does not protrude from the diaphragmatic dome. Image number 1 has a too flattened profile of the superior cardiac margin. (3) Number 2, for the cardiac profile is more faithful to the real contours.

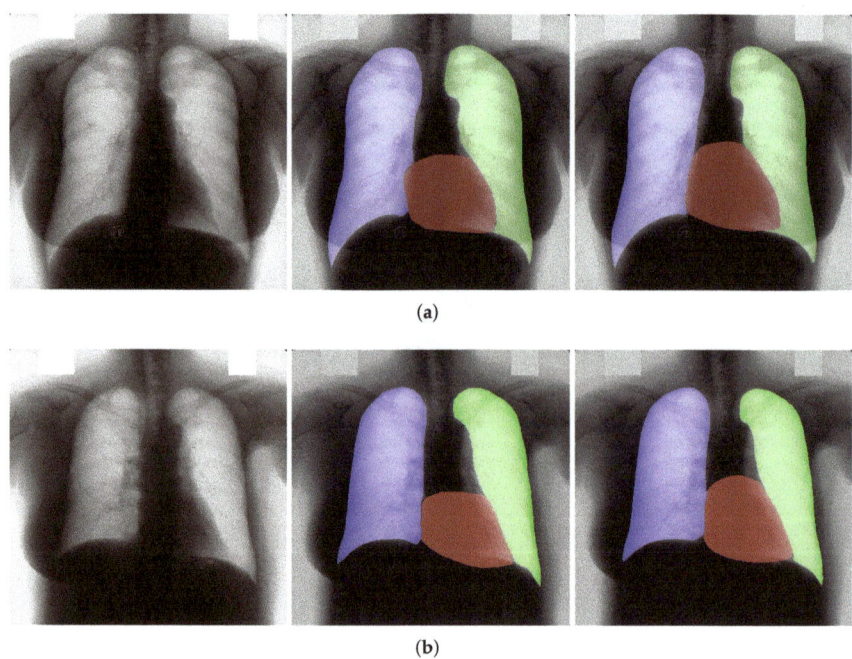

Figure 7. Examples of segmented images for which doctors shared the same opinion. The first column represents the chest X-ray image, while the second and third columns are the target and our predicted segmentation, respectively. (**a**) NODULES001, (**b**) NODULES066.

Furthermore, they reported conflicting opinions or decided not to give a preference with respect to the examples in Figure 8. When they agreed, they generally found different reasons for choosing one segmentation over the other. With respect to Figure 8a the comments reported by the experts were: (1) I prefer not to indicate any options because the heart image is completely subverted; (2) Segmentation number 2 is better, even if it is complicated to read because there is a "bottle-shaped" heart. The only thing that can be improved in image 2 is that a small portion of the right side of the heart is lost; (3) Number 1 respects more what could be the real contours of the heart image. Furthermore, for Figure 8b we obtained: (1) I prefer number 2 because the tip of the heart is well placed on the diaphragm and does not let us see that small wedge-shaped image that incorrectly insinuates itself between heart and diaphragm in image 1 and which has no correspondence in the RX; (2) Both are good segmentations. Both have small problems, for example, in segmentation 1 a small portion of the tip (bottom right of the image) of the heart is missing, in segmentation 2 a part of the outflow cone (the "upper" part of the heart) is missing. It is difficult to choose, probably better number 1 because of the heart; (3) Number 2 because number 1 canal probably exceeds the real dimensions of the cardiac image, including part of the other mediastinal structures.

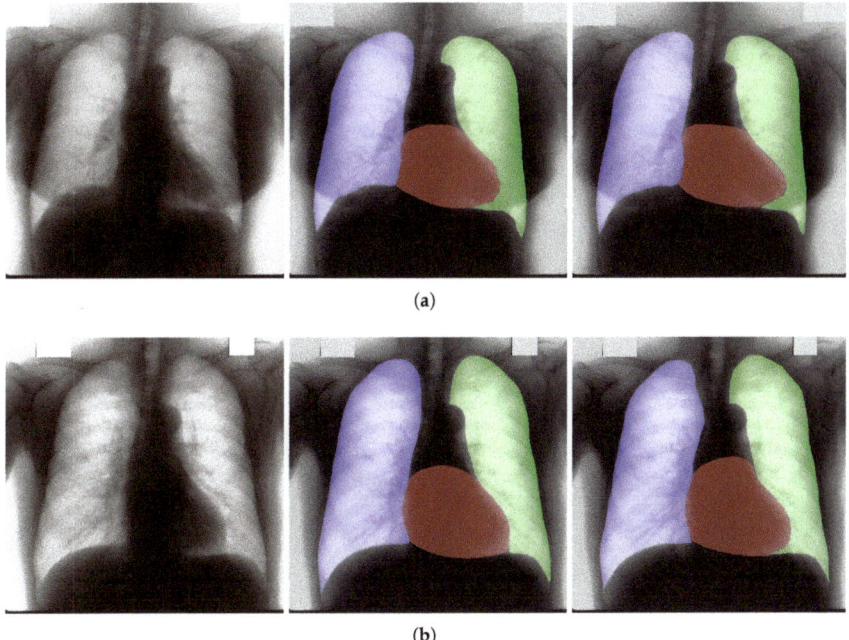

Figure 8. Examples of segmented images for which doctors gave conflicting opinions. The first column represents the chest X-ray image, while the second and third columns are the target and our predicted segmentations, respectively. (**a**) NODULES014, (**b**) NODULES015.

These different evaluations, albeit limited by the small number of examined images, confirm the difficulty of segmenting CXRs, a difficulty that is likely to be more evident in the case of the images selected for our quality analysis, which were chosen based on the large error produced by the segmentation algorithm.

5. Conclusions

In this paper, we have proposed a multi-stage method based on GANs to generate multi-organ segmentation of chest X-ray images. Unlike existing image generation algorithms, in the proposed approach, generation occurs in three stages, starting with "dots", which represent anatomical parts, and initially involves low-resolution images. After the first step, the resolution is increased to translate "dots" into label maps. We performed this step with Pix2PixHD, thus making the information grow and obtaining the labels for each anatomical part taken into consideration. Finally, Pix2PixHD is also used for translating the label maps into the corresponding chest X-ray images. The usefulness of our method was demonstrated especially when there were few images in the training set, an affordable problem thanks to the multi-stage nature of the approach.

It is worth observing that our method can be employed for any type of image, not exclusively medical ones, while synthetic and real images can concur in solving the segmentation problem (being used for pre-training and for fine-tuning the segmentation network, respectively), with a significant increase in performance. As a matter of future research, the proposed approach will be extended to other, more complex domains, such as that of natural images.

Author Contributions: Conceptualization, G.C. and P.A.; methodology, G.C. and P.A.; software, G.C. and P.A.; validation, G.C., P.A., T.M., M.B. and F.S.; formal analysis, G.C. and P.A.; investigation, G.C.; resources, P.A., M.B. and F.S.; data curation, G.C.; writing—original draft preparation, G.C.; writing—review and editing, G.C., P.A., T.M., M.B. and F.S.; visualization, G.C., P.A., T.M., M.B. and F.S.; supervision, M.B. and F.S.; project administration, M.B. and F.S.; funding acquisition, G.C., M.B. and F.S. All authors have read and agreed to the published version of the manuscript.

Funding: The APC was funded by Università degli Studi di Firenze.

Institutional Review Board Statement: Not applicable.

Informed Consent Statement: Not applicable.

Data Availability Statement: http://db.jsrt.or.jp/eng.php.

Acknowledgments: In addition to Tommaso Mazzierli, who is one of the authors of this work, we would like to thank Gabriella Gaudino and Valentina Vellucci for their contribution in the analysis of the segmentations.

Conflicts of Interest: The authors declare no conflict of interest.

References

1. Mettler, F.A., Jr.; Huda, W.; Yoshizumi, T.T.; Mahesh, M. Effective doses in radiology and diagnostic nuclear medicine: A catalog. *Radiology* **2008**, *248*, 254–263. [CrossRef] [PubMed]
2. Hussain, E.; Hasan, M.; Rahman, M.A.; Lee, I.; Tamanna, T.; Parvez, M.Z. CoroDet: A deep learning based classification for COVID-19 detection using chest X-ray images. *Chaos Solitons Fractals* **2021**, *142*, 110495. [CrossRef]
3. Ismael, A.M.; Şengür, A. Deep learning approaches for COVID-19 detection based on chest X-ray images. *Expert Syst. Appl.* **2021**, *164*, 114054. [CrossRef]
4. Nayak, S.R.; Nayak, D.R.; Sinha, U.; Arora, V.; Pachori, R.B. Application of deep learning techniques for detection of COVID-19 cases using chest X-ray images: A comprehensive study. *Biomed. Signal Process. Control* **2021**, *64*, 102365. [CrossRef]
5. Bonechi, S.; Bianchini, M.; Bongini, P.; Ciano, G.; Giacomini, G.; Rosai, R.T.; Rossi, A.R.; Andreini, P. Fusion of Visual and Anamnestic Data for the Classification of Skin Lesions with Deep Learning. In *Lecture Notes in Computer Science*; Cristani, M., Prati, A., Lanz, O., Messelodi, S., Sebe, N., Eds.; Springer: Berlin/Heidelberg, Germany, 2019; Volume 11808, pp. 211–219.
6. Van Ginneken, B.; Stegmann, M.B.; Loog, M. Segmentation of anatomical structures in chest radiographs using supervised methods: A comparative study on a public database. *Med. Image Anal.* **2006**, *10*, 19–40. [PubMed]
7. Qin, C.; Yao, D.; Shi, Y.; Song, Z. Computer-aided detection in chest radiography based on artificial intelligence: A survey. *Biomed. Eng. Online* **2018**, *17*, 1–23. [CrossRef] [PubMed]
8. Teixeira, L.O.; Pereira, R.M.; Bertolini, D.; Oliveira, L.S.; Nanni, L.; Cavalcanti, G.D.; Costa, Y.M. Impact of lung segmentation on the diagnosis and explanation of COVID-19 in chest X-ray images. *arXiv* **2020**, arXiv:2009.09780.
9. Long, J.; Shelhamer, E.; Darrell, T. Fully convolutional networks for semantic segmentation. In Proceedings of the IEEE Conference on Computer Vision and Pattern Recognition, Boston, MA, USA, 7–12 June 2015; pp. 3431–3440.
10. Chen, L.C.; Papandreou, G.; Kokkinos, I.; Murphy, K.; Yuille, A.L. Deeplab: Semantic image segmentation with deep convolutional nets, atrous convolution, and fully connected crfs. *IEEE Trans. Pattern Anal. Mach. Intell.* **2017**, *40*, 834–848. [CrossRef]
11. Zhao, H.; Shi, J.; Qi, X.; Wang, X.; Jia, J. Pyramid scene parsing network. In Proceedings of the IEEE Conference on Computer Vision and Pattern Recognition, Honolulu, HI, USA, 21–26 July 2017; pp. 2881–2890.
12. Goodfellow, I.; Pouget-Abadie, J.; Mirza, M.; Xu, B.; Warde-Farley, D.; Ozair, S.; Courville, A.; Bengio, Y. Generative adversarial nets. *Adv. Neural Inf. Process. Syst.* **2014**, *27*, 2672–2680.
13. Karras, T.; Aila, T.; Laine, S.; Lehtinen, J. Progressive growing of gans for improved quality, stability, and variation. *arXiv* **2017**, arXiv:1710.10196.
14. Wang, T.C.; Liu, M.Y.; Zhu, J.Y.; Tao, A.; Kautz, J.; Catanzaro, B. High-resolution image synthesis and semantic manipulation with conditional gans. In Proceedings of the IEEE Conference on Computer Vision and Pattern Recognition, Salt Lake City, UT, USA, 18–22 June 2018; pp. 8798–8807.
15. Vapnik, V.N. *Statistical Learning Theory*; Wiley-Interscience: Hoboken, NJ, USA, 1998.
16. Neyshabur, B.; Bhojanapalli, S.; Mcallester, D.; Srebro, N. Exploring Generalization in Deep Learning. *Adv. Neural Inf. Process. Syst.* **2017**, *30*, 5947–5956.
17. Kawaguchi, K.; Kaelbling, L.P.; Bengio, Y. Generalization in Deep Learning. *arXiv* **2017**, arXiv:1710.05468.
18. Bonechi, S.; Bianchini, M.; Scarselli, F.; Andreini, P. Weak supervision for generating pixel–level annotations in scene text segmentation. *Pattern Recognit. Lett.* **2020**, *138*, 1–7. [CrossRef]

19. Andreini, P.; Bonechi, S.; Bianchini, M.; Mecocci, A.; Scarselli, F.; Sodi, A. A two stage gan for high resolution retinal image generation and segmentation. *arXiv* **2019**, arXiv:1907.12296.
20. Andreini, P.; Bonechi, S.; Bianchini, M.; Mecocci, A.; Scarselli, F. Image generation by GAN and style transfer for agar plate image segmentation. *Comput. Methods Programs Biomed.* **2020**, *184*, 105268. [CrossRef] [PubMed]
21. Andreini, P.; Bonechi, S.; Bianchini, M.; Mecocci, A.; Scarselli, F. A Deep Learning Approach to Bacterial Colony Segmentation. In *Artificial Neural Networks and Machine Learning—ICANN 2018*; Kůrková, V., Manolopoulos, Y., Hammer, B., Iliadis, L., Maglogiannis, I., Eds.; Springer International Publishing: Cham, Switzerland, 2018; pp. 522–533.
22. Odena, A.; Olah, C.; Shlens, J. Conditional image synthesis with auxiliary classifier gans. In Proceedings of the International Conference on Machine Learning, Sydney, NSW, Australia, 6–11 August 2017; pp. 2642–2651.
23. Karras, T.; Laine, S.; Aila, T. A style-based generator architecture for generative adversarial networks. In Proceedings of the IEEE/CVF Conference on Computer Vision and Pattern Recognition, Long Beach, CA, USA, 16–20 June 2019; pp. 4401–4410.
24. Karras, T.; Laine, S.; Aittala, M.; Hellsten, J.; Lehtinen, J.; Aila, T. Analyzing and improving the image quality of stylegan. In Proceedings of the IEEE/CVF Conference on Computer Vision and Pattern Recognition, Seattle, WA, USA, 13–19 June 2020; pp. 8110–8119.
25. Ledig, C.; Theis, L.; Huszár, F.; Caballero, J.; Cunningham, A.; Acosta, A.; Aitken, A.; Tejani, A.; Totz, J.; Wang, Z.; et al. Photo-realistic single image super-resolution using a generative adversarial network. In Proceedings of the IEEE Conference on Computer Vision and Pattern Recognition, Honolulu, HI, USA, 22–25 July 2017; pp. 4681–4690.
26. Pathak, D.; Krahenbuhl, P.; Donahue, J.; Darrell, T.; Efros, A.A. Context encoders: Feature learning by inpainting. In Proceedings of the IEEE Conference on Computer Vision and Pattern Recognition, Las Vegas, NV, USA, 27–30 June 2016; pp. 2536–2544.
27. Gatys, L.A.; Ecker, A.S.; Bethge, M. A neural algorithm of artistic style. *arXiv* **2015**, arXiv:1508.06576.
28. Liu, M.Y.; Breuel, T.; Kautz, J. Unsupervised image-to-image translation networks. *arXiv* **2017**, arXiv:1703.00848.
29. Liu, M.Y.; Tuzel, O. Coupled generative adversarial networks. *arXiv* **2016**, arXiv:1606.07536.
30. Yi, Z.; Zhang, H.; Tan, P.; Gong, M. Dualgan: Unsupervised dual learning for image-to-image translation. In Proceedings of the IEEE International Conference on Computer Vision, Venice, Italy, 22–29 October 2017; pp. 2849–2857.
31. Zhu, J.Y.; Park, T.; Isola, P.; Efros, A.A. Unpaired image-to-image translation using cycle-consistent adversarial networks. In Proceedings of the IEEE International Conference on Computer Vision, Venice, Italy, 22–29 October 2017; pp. 2223–2232.
32. Isola, P.; Zhu, J.Y.; Zhou, T.; Efros, A.A. Image-to-image translation with conditional adversarial networks. In Proceedings of the IEEE Conference on Computer Vision and Pattern Recognition, Honolulu, HI, USA, 21–26 July 2017; pp. 1125–1134.
33. Chen, Q.; Koltun, V. Photographic image synthesis with cascaded refinement networks. In Proceedings of the IEEE International Conference on Computer Vision, Venice, Italy, 22–29 October 2017; pp. 1511–1520.
34. Mirza, M.; Osindero, S. Conditional generative adversarial nets. *arXiv* **2014**, arXiv:1411.1784.
35. Zhu, J.Y.; Zhang, R.; Pathak, D.; Darrell, T.; Efros, A.A.; Wang, O.; Shechtman, E. Toward multimodal image-to-image translation. *arXiv* **2017**, arXiv:1711.11586.
36. Qi, X.; Chen, Q.; Jia, J.; Koltun, V. Semi-parametric image synthesis. In Proceedings of the IEEE Conference on Computer Vision and Pattern Recognition, Salt Lake City, UT, USA, 18–22 June 2018; pp. 8808–8816.
37. Park, T.; Liu, M.Y.; Wang, T.C.; Zhu, J.Y. Semantic image synthesis with spatially-adaptive normalization. In Proceedings of the IEEE/CVF Conference on Computer Vision and Pattern Recognition, Long Beach, CA, USA, 16–20 June 2019; pp. 2337–2346.
38. Sun, L.; Wang, J.; Ding, X.; Huang, Y.; Paisley, J. An adversarial learning approach to medical image synthesis for lesion removal. *arXiv* **2018**, arXiv:1810.10850.
39. Chen, X.; Konukoglu, E. Unsupervised detection of lesions in brain mri using constrained adversarial auto-encoders. *arXiv* **2018**, arXiv:1806.04972.
40. Schlegl, T.; Seeböck, P.; Waldstein, S.M.; Schmidt-Erfurth, U.; Langs, G. Unsupervised Anomaly Detection with Generative Adversarial Networks to Guide Marker Discovery. *arXiv* **2017**, arXiv:1703.05921.
41. Zhang, X.; Jian, W.; Chen, Y.; Yang, S. Deform-GAN:An Unsupervised Learning Model for Deformable Registration. *arXiv* **2020**, arXiv:2002.11430.
42. Fan, J.; Cao, X.; Xue, Z.; Yap, P.; Shen, D. Adversarial Similarity Network for Evaluating Image Alignment in Deep Learning Based Registration. In Proceedings of the Medical Image Computing and Computer Assisted Intervention—MICCAI 2018—21st International Conference, Granada, Spain, 16–20 September 2018; Lecture Notes in Computer Science; Frangi, A.F., Schnabel, J.A., Davatzikos, C., Alberola-López, C., Fichtinger, G., Eds.; Springer: Berlin/Heidelberg, Germany, 2018; Proceedings, Part I, Volume 11070, pp. 739–746._83. [CrossRef]
43. Tanner, C.; Ozdemir, F.; Profanter, R.; Vishnevsky, V.; Konukoglu, E.; Goksel, O. Generative Adversarial Networks for MR-CT Deformable Image Registration. *arXiv* **2018**, arXiv:1807.07349.
44. Yi, X.; Walia, E.; Babyn, P. Unsupervised and semi-supervised learning with categorical generative adversarial networks assisted by wasserstein distance for dermoscopy image classification. *arXiv* **2018**, arXiv:1804.03700.
45. Madani, A.; Moradi, M.; Karargyris, A.; Syeda-Mahmood, T. Semi-supervised learning with generative adversarial networks for chest X-ray classification with ability of data domain adaptation. In Proceedings of the 2018 IEEE 15th International Symposium on Biomedical Imaging (ISBI 2018), Washington, DC, USA, 4–7 April 2018; pp. 1038–1042.
46. Lecouat, B.; Chang, K.; Foo, C.S.; Unnikrishnan, B.; Brown, J.M.; Zenati, H.; Beers, A.; Chandrasekhar, V.; Kalpathy-Cramer, J.; Krishnaswamy, P. Semi-Supervised Deep Learning for Abnormality Classification in Retinal Images. *arXiv* **2018**, arXiv:1812.07832.

47. Li, Y.; Shen, L. cC-GAN: A robust transfer-learning framework for HEp-2 specimen image segmentation. *IEEE Access* **2018**, *6*, 14048–14058. [CrossRef]
48. Xue, Y.; Xu, T.; Zhang, H.; Long, L.R.; Huang, X. Segan: Adversarial network with multi-scale l 1 loss for medical image segmentation. *Neuroinformatics* **2018**, *16*, 383–392. [CrossRef]
49. Frid-Adar, M.; Diamant, I.; Klang, E.; Amitai, M.; Goldberger, J.; Greenspan, H. GAN-based synthetic medical image augmentation for increased CNN performance in liver lesion classification. *Neurocomputing* **2018**, *321*, 321–331. [CrossRef]
50. Hu, B.; Tang, Y.; Eric, I.; Chang, C.; Fan, Y.; Lai, M.; Xu, Y. Unsupervised learning for cell-level visual representation in histopathology images with generative adversarial networks. *IEEE J. Biomed. Health Inform.* **2018**, *23*, 1316–1328. [CrossRef] [PubMed]
51. Srivastav, D.; Bajpai, A.; Srivastava, P. Improved Classification for Pneumonia Detection using Transfer Learning with GAN based Synthetic Image Augmentation. In Proceedings of the 2021 11th International Conference on Cloud Computing, Data Science & Engineering (Confluence), Noida, India, 28–29 January 2021; pp. 433–437.
52. Candemir, S.; Jaeger, S.; Palaniappan, K.; Musco, J.P.; Singh, R.K.; Xue, Z.; Karargyris, A.; Antani, S.; Thoma, G.; McDonald, C.J. Lung segmentation in chest radiographs using anatomical atlases with nonrigid registration. *IEEE Trans. Med Imaging* **2013**, *33*, 577–590. [CrossRef] [PubMed]
53. Boykov, Y.; Funka-Lea, G. Graph cuts and efficient ND image segmentation. *Int. J. Comput. Vis.* **2006**, *70*, 109–131. [CrossRef]
54. Candemir, S.; Akgül, Y.S. Statistical significance based graph cut regularization for medical image segmentation. *Turk. J. Electr. Eng. Comput. Sci.* **2011**, *19*, 957–972.
55. Boykov, Y.; Jolly, M. Interactive graph cuts for optimal boundary and region segmentation of objects in nd images. In Proceedings of the Eighth IEEE International Conference on Computer Vision, Vancouver, BC, Canada, 7–14 July 2001; pp. 105–112.
56. Shao, Y.; Gao, Y.; Guo, Y.; Shi, Y.; Yang, X.; Shen, D. Hierarchical lung field segmentation with joint shape and appearance sparse learning. *IEEE Trans. Med Imaging* **2014**, *33*, 1761–1780. [CrossRef]
57. Ibragimov, B.; Likar, B.; Pernuš, F.; Vrtovec, T. Accurate landmark-based segmentation by incorporating landmark misdetections. In Proceedings of the 2016 IEEE 13th International Symposium on Biomedical Imaging (ISBI), Prague, Czech Republic, 13–16 April 2016; pp. 1072–1075.
58. Novikov, A.A.; Lenis, D.; Major, D.; Hladůvka, J.; Wimmer, M.; Bühler, K. Fully Convolutional Architectures for Multiclass Segmentation in Chest Radiographs. *IEEE Trans. Med Imaging* **2018**, *37*, 1865–1876. [CrossRef]
59. Ronneberger, O.; Fischer, P.; Brox, T. U-net: Convolutional networks for biomedical image segmentation. In Proceedings of the International Conference on Medical Image Computing and Computer-Assisted Intervention, Munich, Germany, 5–9 October 2015; pp. 234–241.
60. Wang, C. Segmentation of multiple structures in chest radiographs using multi-task fully convolutional networks. In Proceedings of the Scandinavian Conference on Image Analysis, Tromsø, Norway, 12–14 June 2017; pp. 282–289.
61. Oliveira, H.; dos Santos, J. Deep transfer learning for segmentation of anatomical structures in chest radiographs. In Proceedings of the 2018 31st SIBGRAPI Conference on Graphics, Patterns and Images (SIBGRAPI), Paraná, Brazil, 29 October–1 November 2018; pp. 204–211.
62. Islam, J.; Zhang, Y. Towards robust lung segmentation in chest radiographs with deep learning. *arXiv* **2018**, arXiv:1811.12638.
63. Dai, W.; Dong, N.; Wang, Z.; Liang, X.; Zhang, H.; Xing, E.P. Scan: Structure correcting adversarial network for organ segmentation in chest x-rays. In *Deep Learning in Medical Image Analysis and Multimodal Learning for Clinical Decision Support*; Springer: Berlin/Heidelberg, Germany, 2018; pp. 263–273.
64. He, K.; Zhang, X.; Ren, S.; Sun, J. Deep residual learning for image recognition. In Proceedings of the IEEE Conference on Computer Vision and Pattern Recognition, Las Vegas, NV, USA, 27–30 June 2016; pp. 770–778.
65. Papandreou, G.; Kokkinos, I.; Savalle, P.A. Untangling local and global deformations in deep convolutional networks for image classification and sliding window detection. *arXiv* **2014**, arXiv:1412.0296.
66. Kingma, D.P.; Ba, J. Adam: A method for stochastic optimization. *arXiv* **2014**, arXiv:1412.6980.
67. Shiraishi, J.; Katsuragawa, S.; Ikezoe, J.; Matsumoto, T.; Kobayashi, T.; Komatsu, K.i.; Matsui, M.; Fujita, H.; Kodera, Y.; Doi, K. Development of a digital image database for chest radiographs with and without a lung nodule: Receiver operating characteristic analysis of radiologists' detection of pulmonary nodules. *Am. J. Roentgenol.* **2000**, *174*, 71–74. [CrossRef]

MDPI
St. Alban-Anlage 66
4052 Basel
Switzerland
Tel. +41 61 683 77 34
Fax +41 61 302 89 18
www.mdpi.com

Mathematics Editorial Office
E-mail: mathematics@mdpi.com
www.mdpi.com/journal/mathematics

www.ingramcontent.com/pod-product-compliance
Lightning Source LLC
LaVergne TN
LVHW070544100526
838202LV00012B/372